CISM Courses and Lectures

The series presents lecture notes, monographs, edited works and proceedings in the field of Mechanics, Engineering, Computer Science and Applied Mathematics.
Purpose of the series is to make known in the international scientific and technical community results obtained in some of the activities organized by CISM, the International Centre for Mechanical Sciences.

International Centre for Mechanical Sciences

Courses and Lectures Vol. 561

For further volumes:
www.springer.com/series/76

Gernot Beer · Stéphane Bordas
Editors

Isogeometric Methods for Numerical Simulation

Springer

Editors

Gernot Beer
Technical Univ. of Graz, Graz, Austria

Stéphane Bordas
Cardiff University, Cardiff, Great Britain

ISSN 0254-1971
ISBN 978-3-7091-4881-5   ISBN 978-3-7091-1843-6 (eBook)
DOI 10.1007/978-3-7091-1843-6
Springer Wien Heidelberg New York Dordrecht London

Softcover re-print of the Hardcover 1st edition 2015

All contributions have been typeset by the authors
Printed in Italy

Printed on acid-free paper

Springer is part of Springer Science+Business Media (www.springer.com)

PREFACE

The initial idea of isogeometric methods is due to Tom Hughes, who proposed a tighter connection between computer aided design (CAD) and simulation. Practically, most simulations are performed based on some sort of model of the geometry constructed within a CAD framework. Subsequently, this CAD geometry is "meshed" so that the problem can be solved using the finite element method (FEM) or the Boundary Element Method (BEM). Based on the solution of the problem, engineers monitor quantities of interest, e.g. the maximum stress or displacement in a structure to decide upon the suitability of the chosen geometry for the intended use of the object. If this initial geometry must be modified, the corresponding mesh must usually be at least partly discarded, and a new mesh constructed, for the new geometry. This process must be repeated as many times as is required to converge to an allowable solution.

It thus appears that the practice of using the CAD data to generate a series of meshes for simulations is cumbersome, time consuming and prone to errors. Anyone that has performed serious simulation work is aware of the effort required to produce a mesh of suitable quality and of the pitfalls associated with mesh generation. Although many excellent mesh generators are available, to generate meshes from CAD data, there is always the possibility that for awkward geometries, elements with bad aspect ratios are encountered and the mesh has to be repaired to yield acceptable results. Also, in addition to the (discretization) error introduced by approximating the unknown solution, one has to control the (geometrical) error introduced when approximating the geometry by the mesh. For some applications in shell theory, electromagnetics for example and many others, an accurate description of the geometry is paramount. Any method able to reduce both discretization and geometrical errors is expected to be welcomed by practitioners. It was the very goal of this course on Isogeometric Analysis, addressed to newcomers to the field, to simplify the transition from the standard "CAD-Analysis" pipeline to the "Isogeometric" paradigm where meshing is either completely suppressed (in Isogeometric Boundary Element Methods) or greatly simplified (in Isogeometric Finite Element, Collocation and Meshless Methods).

Although the initial intention of combining CAD and simulation was no doubt to avoid mesh generation by using the CAD data directly for the simulation, it became soon clear that there was an added value to this approach. The functions used by CAD programs to describe the geometry (NURBS) also happened to have significant advantages in approximating the unknown fields. Early work reported in the first book on the topic[1] *indicated that with these functions much better results could be obtained with a smaller number of degrees of freedom. This appealing property is partly due to NURBS' improved continuity properties and are amenable to efficient hpk-refinement strategies.*

The aim of the course "Isogeometric Methods for numerical simulation" was to give an in depth overview of the fundamentals behind this technology and to present some of its most recent applications to a range of problems from fluid to fracture mechanics simulations. This book contains some of the lecture material presented at the course.

The book starts with an introduction by B. Jüttler to B-splines and NURBS. The next chapter by V.P. Nguyen and S. Bordas deals with the implementation of isogeometric methods in structural mechanics using Finite Element and eXtended Finite Element Methods, which can be used to model cracks and delamination in composite structures. This chapter also refers to alternative methods to isogeometric analysis, in particular when the field and the geometrical approximations are to be kept independent. The next chapter covers the implementation of isogeometric techniques into the BEM. It can be seen that the BEM is an ideal companion to CAD because both use a surface representation. Here G. Beer and B. Marussig present the basic idea of a geometry independent approximation of field variables for the first time. The next chapter by A. Reali and T.J.R. Hughes provides a concise introduction to isogeometric collocation methods and proposes a brief review of some of the most important results obtained so far in that context. In the last chapter some new concepts for the description of complex surfaces are presented. D. Thomas and M. Scott explain how such surfaces can be described and simulated with few control points using T-spline technology. The course also covered fluid-structure interaction, through most inspiring lectures by Y. Bazilevs who was, unfortunately, unable to contribute to this book.

---

[1] J.A. Cottrell, T.J.R. Huges and Y. Bazilevs *Isogeometric Analysis*, J. Wiley 2009

The course was very well attended and we would like to thank all participants, in particular those coming from industry, for their interest and for the many fruitful discussions which ensued. Thanks are also due to the lecturers for taking time within busy schedules to prepare and present excellent lectures. We hope that the course and this book will inspire many to work in this exciting and continuously expanding area.

Last but not least, we would like to thank the people at Springer for their assistance and patience.

Gernot Beer and Stéphane Bordas

CONTENTS

A Primer on Splines and NURBS for Isogeometric Analysis

Bert Jüttler

Institute of Applied Geometry, Johannes Kepler University, Linz, Austria.
Email: bert.juettler@jku.at

Abstract This chapter collects fundamental facts concerning theory and algorithms for rational spline functions. In addition, it addresses fundamental techniques for creating domain parameterizations for isogeometric analysis. All presented facts are covered in more detail by the the rich literature on Computer Aided Geometric Design, e.g., (Fa02; FHK02; HL96; PT95; PBP02),

1 Spline functions as piecewise polynomials

We define univariate spline functions as piecewise polynomial functions with prescribed smoothness. In addition we introduce a basis for the linear space formed by them.

1.1 The space of spline functions

We consider

- a given polynomial degree d and a positive integer m, which satisfies $m \geq 2d$,
- a given monotonically increasing ($\tau_i \leq \tau_{i+1}$) sequence of real numbers, the *knot sequence*

$$\mathcal{K} = (\tau_0, \ldots, \tau_{m+1}),$$

  satisfying

$$\tau_i < \tau_{i+d} \text{ for } i = 1, \ldots, m-d \text{ and } \tau_d < \tau_{d+1}, \quad \tau_{m-d} < \tau_{m-d+1}$$

- and the *domain*

$$D = (\tau_d, \tau_{m-d+1}).$$

The following definition describes how one encodes the *smoothness* of piecewise polynomial functions of a given fixed degree d with the help of the knot vector:

G. Beer, S. Bordas (Eds.), *Isogeometric Methods for Numerical Simulation*,
CISM International Centre for Mechanical Sciences
DOI 10.1007/ 978-3-7091-1843-6_1 © CISM Udine 2015

Definition 1.1. A real-valued function with domain D is called a *spline function* of degree d with the knots $\mathcal{K}$, if the restriction of the function to each non-empty interval (τ_i, τ_{i+1}) is a polynomial of degree d and the function f is at least $d-k$ times continuously differentiable at each inner knot $\tau_i \in D$ with multiplicity k, i.e., at each knot which satisfies

$$\tau_{i-1} < \tau_i = \ldots = \tau_{i+k-1} < \tau_{i+k} \quad \text{and } \tau_d < \tau_i < \tau_{m+d-1}.$$

The knots define the junction points where the different polynomial segments are pieced together. The multiplicity encodes the required smoothness at these points. Due to the constraint on the multiplicity, the function is globally continuous.

The condition $\tau_i < \tau_{i+m}$ ensures that at most m knots coincide. This condition is not required for the first and for the last knot ($i = 0$ and $i = m - d + 1$).

The location of the first and last d knots has no influence on the definition of the spline space, since these knots are not contained in the interior of the domain. Typically one chooses $d + 1$ fold boundary knots

$$\tau_0 = \ldots = \tau_d \text{ and } \tau_{m-d+1} = \ldots = \tau_{m+1}.$$

The set of spline functions for given knots and degree forms a linear space $\mathbb{S}^d_{\mathcal{K}}$. A simple basis for this space is obtained by collecting basis polynomials and truncated power functions:

Proposition 1.2. *The truncated power basis of the linear space of spline functions of degree d with knots $\mathcal{K}$ consists of*

- *the monomials $x \mapsto x^i$ for $i = 0, \ldots, d$ and*
- *the k truncated power functions*

$$x \mapsto (x - \tau_i)_+^{d-k+1}, \ldots, x \mapsto (x - \tau_i)_+^d$$

for each inner knot τ_i of multiplicity k, where

$$(x)_+ = \frac{1}{2}(x + |x|) = \begin{cases} x \text{ if } x > 0 \\ 0 \text{ otherwise} \end{cases}.$$

Consequently, the *dimension* of the space of spline functions is equal to

$$\dim \mathbb{S}^d_{\mathcal{K}} = m - d + 1.$$

This basis is not very useful in practice, since it consists of basis functions with global support. Moreover, the functions do not possess properties such as non-negativity or the partition of unity property.

1.2 B-splines

The B-splines ("Basis splines") form another basis for the space of spline functions. For a given knot sequence $\mathcal{K}$, these functions are defined by a recurrence relation.

Definition 1.3. The *B-splines of degree 0* are the characteristic functions of the half-open knot spans,

$$B_i^0(x) = \begin{cases} 1 \text{ if } x \in [\tau_i, \tau_{i+1}) \\ 0 \text{ otherwise} \end{cases}, \quad i = 0, \ldots, m.$$

The *B-splines of degree* k are obtained recursively from those of degree $k-1$,

$$B_i^k(x) = \frac{x - \tau_i}{\tau_{i+k} - \tau_i} B_i^{k-1}(x) + \frac{\tau_{i+k+1} - x}{\tau_{i+k+1} - \tau_{i+1}} B_{i+1}^{k-1}(x), \quad i = 0, \ldots, m-k.$$

In the case of multiple knots one may encounter situations where some of the denominators vanish. In this situation, however, also the B-spline B_j^{k-1} in this term vanishes and the term is considered to be 0.

Sometimes one uses the *order* $k+1$ instead of the degree k to identify the recursion level.

The B-splines form a convex partition of unity. More precisely, these functions take only non-negative values and their sum is equal to 1 in the domain. Each B-spline B_i^n has the support $[\tau_i, \tau_{i+d+1}]$ and it is fully determined by the $d+2$ knots within its support.

The B-splines form a basis of the spline space:

Proposition 1.4. *Each spline function possesses a representation of the form*

$$s(x) = \sum_{i=0}^{m-d} B_i^d(x) c_i \quad x \in D,$$

with certain coefficients $c_i \in \mathbb{R}$.

2 Blossoming

Blossoming (Ra89) is a technique that makes it possible to derive the algorithms for spline functions in B-spline representation in a particularly elegant way.

2.1 The blossom of a polynomial

We consider a polynomial $p : \mathbb{R} \to \mathbb{R}$ of degree d.

Definition 2.1. The mapping $P : \mathbb{R}^d \to \mathbb{R}$ is called a *blossom* (or polar form) of p, if it possesses the following three properties.

1. P is multi-affine, i.e., it is a linear polynomial in each of its d arguments.
2. P is totally symmetric, i.e., arbitrary permutations of its d arguments leave the value of P unchanged.
3. The restriction of P to the diagonal is the polynomial p,

$$P(\underbrace{t, \ldots, t}_{d \text{ times}}) = p(t).$$

First we verify the existence of the blossom.

Proposition 2.2. *The blossom of a polynomial exists.*

Proof. We consider a polynomial of degree d in its represented with respected to the shifted monomials $(x - x_0)^i$, $i = 0, \ldots, d$,

$$p(x) = \sum_{i=0}^{d} c_i (x - x_0)^i.$$

This representation will be useful for proving the theorem concerning the contact of two polynomials below. The blossom of this polynomial takes the form

$$P(x_1, \ldots, x_d) = c_0 + \sum_{i=1}^{d} c_i \frac{1}{\binom{d}{i}} \left(\sum_{\{j_1, \ldots, j_i\} \in C_i} \prod_{k=1}^{i} (x_{j_k} - x_0) \right) \tag{1}$$

where the set C_i, which is considered in the summation, consists of all subsets $\{j_1, \ldots, j_i\}$ of the index set $\{1, \ldots, d\}$ that possess exactly i elements. Indeed, it is easy to see this blossom satisfies the three properties (a-c). □

The expression in Eq. (1) is called the *Taylor form* of the blossom since it is related to the Taylor expansion at a point x_0.

The generalized algorithm of de Boor allows to compute any value of the blossom from a particular set of $d + 1$ values. More precisely, the input consists of $d + 1$ values of the blossom

$$[P(k_{i+1}, \ldots, k_{i+d})]_{i=0,\ldots,d}, \tag{2}$$

associated with adjacent d-tuples of nodes k_i, where the $2d$ nodes $k_1, \ldots k_{2d}$ are monotonically increasing and satisfy $k_i < k_{i+d}$, i.e., at most d nodes take the same value.

Definition 2.3. Given the values (2) of the blossom, the *generalized algorithm of de Boor* evaluates the blossom at an arbitrary point $P(x_1, \ldots, x_d)$. The algorithm uses the relation

$$\begin{aligned}&P(k_{j+1}, \ldots, k_{j+r}, x_1, \ldots, x_{d-r}) = \\&= \alpha(k_j, k_{j+r+1}, x_{d-r})\, P(k_j, \ldots, k_{j+r}, x_1, \ldots, x_{d-r-1}) + \\&+ [1 - \alpha(k_j, k_{j+r+1}, x_{d-r})]\, P(k_{j+1}, \ldots, k_{j+r+1}, x_1, \ldots, x_{d-r-1})\end{aligned}$$

$r = d-1, (-1), 0$ and $j = d-r, \ldots, d$ to compute new values of the blossom from known ones. Here the numbers $\alpha(\ldots)$ and $1-\alpha(\ldots)$ are the barycentric coordinates of x_{d-r} with respect to the interval $[k_j, k_{j+r+1}]$,

$$\alpha(k_j, k_{j+r+1}, x_{d-r}) = \frac{k_{j+r+1} - x_{d-r}}{k_{j+r+1} - k_j}.$$

Eventually, for $r = 0$ and $j = d$, the algorithm arrives at final result.

The recurrence relation follows from the first two properties of the blossom (symmetry and multi-affine property). The output is computed by applying the recurrence relation in order to replace the given node values with the real numbers x_i. This gives a triangular scheme.

Two special cases of de Boor's algorithm are of particular interest.

- **Algorithm of de Boor:** The d coordinates of the evaluation point take the same value,
$$x_1 = \ldots = x_d = x.$$
For this choice of the evaluation point, the algorithm evaluates the polynomial at x, due to the diagonal property of the blossom.
- **Algorithm of de Casteljau:** The nodes take the special form
$$k_1 = \ldots = k_d = 0, \quad k_{d+1} = \ldots = k_{2d} = 1$$
and also coordinates of the evaluation point are again identical. In this special situation, the algorithm simplifies to the Algorithm of de Casteljau, which evaluates a polynomial in Bernstein-Bézier representation.

The blossom of a polynomial is unique. The proof of this fact can be obtained using de Boor's algorithm.

The *contact* (C^r joint) of two polynomial functions at a given can be characterized using their blossoms as follows.

Theorem 2.4. *The following two statements are equivalent.*

(i) The two polynomials f and g have a C^r joint at x_0, i.e.,

$$(\frac{\mathrm{d}^i}{\mathrm{d}x^i}f)(x_0) = (\frac{\mathrm{d}^i}{\mathrm{d}x^i}g)(x_0), \quad i = 0, \ldots, r.$$

(ii) The blossoms satisfy

$$\forall x_1, \ldots, x_r : F(x_0, \ldots, x_0, x_1, \ldots, x_r) = G(x_0, \ldots, x_0, x_1, \ldots, x_r)$$

where x_0 is repeated $d - r$ times.

This observation can be proved easily with the help of the Taylor form (1) of the blossom.

2.2 Spline functions in B-spline form

We recall the definition of a spline functions and rewrite it with the help of the blossom.

Definition 2.5. A *spline function in B-spline form* is defined as follows. We consider

- a monotonically increasing knot sequence

$$\mathcal{T} = (\tau_1, \ldots, \tau_m),$$

  satisfying $\tau_i < \tau_{i+d}$, and
- set of *control points* (coefficients) associated with d-tuples of consecutive knots,

$$c_i \sim \tau_{i+1}, \ldots, \tau_{i+d}, \quad i = 0, \ldots, m - d.$$

Let p_j be the polynomial of degree d with the blossom P_j satisfying

$$P_j(\tau_{i+1}, \ldots, \tau_{i+d}) = c_i, \quad i = j, \ldots, j + d,$$

where $j = 0, \ldots, m - 2d$. The spline function is defined by

$$s(x) = p_j(x) \quad \text{if} \quad x \in [\tau_{j+d}, \tau_{j+d+1}].$$

Its domain is the interval $D = [\tau_d, \tau_{m-d+1}]$.

It should be noted that each polynomial p_j (and its blossom) can be evaluated with the (generalized) algorithm of de Boor. Indeed, the $2d$ nodes required by this algorithm are the knots $\tau_{j+1}, \ldots, \tau_{j+2d}$, and the control points c_i are exactly the values (2) required by the algorithm.

The previous definition of a spline function (Definition 1.1) uses a knot vector $\mathcal{K}$ with two additional knots on either end of the knot vector. These knots are required if the B-spline recurrence is used, but they do not have any influence and can therefore be omitted.

First we analyze the smoothness of a spline function in B-spline representation.

Proposition 2.6. *If a knot τ_i has multiplicity k, i.e., a knot that satisfies $\tau_{i-1} < \tau_i = \ldots = \tau_{i+k-1} < \tau_{i+k}$, then the spline function s defined in the previous definition is at least C^{d-k} smooth at $x = \tau_i$.*

This fact is implied by the Theorem 2.4 and can be verified with the help of de Boor's algorithm.

Spline functions in B-spline representation possess the convex hull property.

Proposition 2.7. *If $x \in [\tau_{j+d}, \tau_{j+d+1}]$, then*

$$\min\{(c_i)_{i=j,\ldots,j+d}\} \leq s(x) \leq \max\{(c_i)_{i=j,\ldots,j+d}\}.$$

This can be proved by observing that the algorithm of de Boor uses only convex linear combinations when evaluating a spline function in B-spline representation.

2.3 B-splines

Definition 2.8. The spline function with control points

$$c_i = \delta_{ik} = \begin{cases} 1 \text{ if } i = k \\ 0 \text{ otherwise} \end{cases} \tag{3}$$

is called the *B-spline* (Basis-spline) B_k associated with the knot sequence $\mathcal{T}$.

Proposition 2.9. *(Properties of B-splines) The B-splines are non-negative functions. In addition, they satisfy*

$$B_k(x) > 0 \iff x \in (\tau_k, \tau_{k+d+1}).$$

The values of the B-spline B_k do not depend on the knots τ_i with $i < k$ or $i > k + d + 1$. The sum of the B-splines is equal to 1,

$$\sum_{k=0}^{m-d} B_k(x) = 1, \quad x \in D.$$

Note that – in this section – the last and the first B-spline are defined with one knot less than the remaining B-splines, since the knot sequence $\mathcal{T}$ contains two knots less than $\mathcal{K}$.

Each spline function has a representation of the form

$$s(x) = \sum_{i=0}^{m-d} B_i(x)c_i.$$

Moreover, the B-splines B_i and B_i^d (see Definition 1.3) are identical. Consequently, the B-splines B_i form a basis of the spline space $\mathbb{S}_{\mathcal{K}}^d$.

3 Computations with spline functions

As the main advantage, the blossom allows to derive various useful algorithms for computing with B-splines. We consider the evaluation of B-spline coefficients of polynomials and the computation of the derivative of a spline function.

3.1 B-spline coefficients of polynomials

Given a polynomial p of degree d, the B-spline coefficients can be found by evaluating the blossom,

$$c_i = P(\tau_{i+1}, \ldots, \tau_{i+d}), i = 0, \ldots, m - d.$$

In particular, this is true for the linear polynomial $p(x) = x$. The corresponding values of its blossom

$$P(x_1, \ldots, x_d) = \frac{1}{d}(x_1 + \ldots + x_d)$$

are the so-called *Greville points*

$$\gamma_i = \frac{1}{d}(\tau_{i+1} + \ldots + \tau_{i+d}).$$

3.2 Differentiation

Given a spline function in B-spline representation, we derive a B-spline representation of its derivative. We assume that the multiplicity of the inner knots is strictly less than d. First we consider the blossom of the derivative of a polynomial:

Lemma 3.1. *Let P' be the blossom of the derivative p' of a polynomial of degree d. Then, for any y, z, $y \neq z$,*

$$P'(x_1, \ldots, x_{d-1}) = \frac{d}{y - z}[P(x_1, \ldots, x_{d-1}, y) - P(x_1, \ldots, x_{d-1}, z)].$$

Indeed, it is easy to check that P' satisfies the first two properties of the blossom. The third one can be verified by a simple computation.

This result allows us to compute the derivative of a spline function in B-spline representation.

Proposition 3.2. *The derivative of a spline curve has a B-spline representation of degree $d-1$ with the knots*

$$\mathcal{T}' = (\tau_2, \ldots, \tau_{m-1}) = (\tau'_1, \ldots, \tau'_{m-2})$$

and control points c'_i associated with the knots

$$c'_i \sim (\tau'_{i+1}, \ldots, \tau'_{i+d-1}).$$

These control points satisfy

$$c'_i = \frac{c_{i+1} - c_i}{\tau_{i+d+1} - \tau_{i+1}}.$$

3.3 Products and compositions of spline functions

Now we turn our attention to products and compositions of spline functions. The blossom of the product of two polynomials p and q of degree d and e, respectively, is discussed first. The product is a polynomial of degree $d+e$.

Lemma 3.3. *Consider two polynomials p, q of degree d, e with blossoms P, Q. Their product $h = pq$ has the blossom*

$$H(x_1, \ldots, x_{d+e}) = \frac{1}{\binom{d+e}{d}} \sum_{(\{i_1,\ldots,i_d\},\{j_1,\ldots,j_e\})\in C} P(x_{i_1}, \ldots, x_{i_d}) Q(x_{j_1}, \ldots, x_{j_e}).$$

The set C, which is considered in the summation, contains all pairs of disjoint subsets of the index set $\{1, \ldots, d+e\}$, where the first subset contains d elements and the second one contains the remaining e elements.

This lemma can be used to derive formulas and algorithms for computing the control points of products of two spline functions in B-spline form. A similar result can be obtained for the blossom of the composition of the two polynomials. In this case, the composition has the polynomial degree de.

Lemma 3.4. *Consider two polynomials p, q of degree d, e with blossoms P, Q. Their composition $h = p \circ q$ has the blossom*

$$H(x_1, \ldots, x_{de}) = \frac{(e!)^d}{(de)!} \sum_{(\{i_1^{(1)},\ldots,i_e^{(1)}\},\ldots,\{i_1^{(d)},\ldots,i_e^{(d)}\})\in C} P(Q(x_{i_1^{(1)}}, \ldots, x_{i_e^{(1)}}), \ldots, Q(x_{i_e^{(d)}}, \ldots, x_{i_e^{(d)}}))$$

where the summation considers the set C of all d-tuples of disjoint subsets of the index set $\{1, \ldots, de\}$ with e elements each.

As in the case of products, this lemma can be used to derive formulas and algorithms for computing the control points of compositions of two spline functions in B-spline form.

3.4 Knot insertion

The knot insertion is probably the most important algorithm for spline functions in B-spline representation. The B-spline representation is modified by inserting a single knot into the original knot sequence.

The B-spline representation of a given spline function with respect to the modified knot vector $T \cup \{x\}$ (where the new knot x is inserted at the appropriate place) can be computed as follows.

1. Find the interval $[\tau_i, \tau_{i+1})$ which contains the new knot x.
2. Apply the first step of the de Boor algorithm to the following $d+1$ values of the blossom,

$$c_{i-d+j} = P_{i-d}(\tau_{i-d+j+1}, \ldots, \tau_{i+j}), \quad j = 0, \ldots, d.$$

3. The new control points are

$$\begin{aligned} c'_j &= c_j, \quad j = 0, \ldots, i-d, \\ c'_j &= P_{i-d}(\tau_{j+1}, \ldots, \tau_{j+d-1}, x), \quad j = i-d+1, \ldots, i, \\ c'_j &= c_{j-1}, \quad j = i+1, \ldots, m-d+1. \end{aligned}$$

The knot insertion can be used for evaluation at a point x: The knot x is inserted until it reaches multiplicity d. The corresponding control points gives the value of the spline function.

It can also be used for for splitting the function into Bézier segments ("Oslo algorithm"): The existing knots are inserted until they reach the multiplicity d. The coefficients are then the Bernstein-Bézier coefficients of the polynomial segments.

Finally knot insertion can also be used for subdividing B-splines into B-splines with smaller support: The knot insertion is applied to Kronecker data (see (3)).

4 Evaluation techniques for spline functions

We discuss two different approaches for evaluating a spline function.

First we consider *exact* evaluation at a point x. This can be achieved either by using de Boor's algorithm or via the B-spline basis. Evaluation by the de Boor algorithm proceeds as follows:

1. Find the interval $[\tau_i, \tau_{i+1})$ which contains the new knot x.
2. Apply the de Boor algorithm to the following $d+1$ values of the blossom,

$$c_{i-d+j} = P_{i-d}(\tau_{i-d+j+1}, \ldots, \tau_{i+j}), \quad j = 0, \ldots, d.$$

3. Return the value $s(x) = P_{i-d}(x, \ldots, x)$ of the spline function at x.

This evaluation procedure is numerically stable.

Alternatively, the spline function can be evaluated by first evaluating the B-splines and then summing up:

1. Find the interval $[\tau_i, \tau_{i+1})$ that contains x.
2. Evaluate the $d+1$ B-splines

$$B_{i-d+l}(x), \quad l = 0, \ldots, d.$$

3. Evaluate and return the value of the spline function at x,

$$s(x) = \sum_{j=0,\ldots,d} B_{i-d+j}(x) c_{i-d+j}$$

The evaluation of the B-splines in the second step can be performed by using the B-spline recurrence, by using (again) de Boor's algorithm, or by a direct evaluation of pre-computed polynomial segments.

Second, approximate evaluation via knot insertion is available and used frequently in Computer Graphics. It is is based on the control polygon (a piecewise linear function) and knot insertion.

Recall that the Greville points

$$\gamma_i = \frac{1}{d}(\tau_{i+1} + \ldots + \tau_{i+d})$$

are the coefficients of the function $s(x) = x$, i.e.,

$$x = \sum_{i=0}^{m-d} \gamma_i B_i(x)$$

The piecewise linear function with the vertices $(\gamma_i, c_i)_{i=0,\ldots,m-d}$ is called the *control polygon* of the spline function. The control polygon approximates the spline function. If the knot sequence is refined such that the maximum difference of adjacent knots tends to zero, then the control polygon converges to the function. This is due to the fact that the blossom is continuous.

This observation leads to a framework for approximate evaluation:

- Apply knot insertion until the distance between adjacent knots is below a certain threshold / until the control polygon is sufficiently close to a straight line locally.
- Output: Piecewise linear function with vertices (ξ_i, d_i), $s(\xi_i) \approx d_i$.

This techniques allows for simultaneous evaluation of many values, which is especially useful for visualization purposes.

Two special cases lead to uniform refinement rules:

- Bézier subdivision: All knots have multiplicity d, and the newly inserted knots are the midpoints of the intervals and have multiplicity d, too.
- Lane-Riesenfeld subdivision: All knots are single knots and are equidistant, and all knot intervals are halved.

In these two special cases, one obtains uniform refinement rules which are based on normalized binomial coefficients.

5 NURBS objects

Based on the definitions of B-splines, we introduce piecewise rational parametric representations of curves, surfaces, and volumes.

5.1 NURBS curves

We consider four spline functions $w(t), x(t), y(t), z(t)$, which possess the same knots $\mathcal{T}$ and degree d. The coefficients are $c_i^w, c_i^x, c_i^y, c_i^z$. The piecewise rational curve

$$t \mapsto \mathbf{p}(t) = \frac{1}{w(t)} \begin{pmatrix} x(t) \\ y(t) \\ z(t) \end{pmatrix} = \frac{1}{\sum_j B_j(t) c_j^w} \sum_i B_i(t) \begin{pmatrix} c_i^x \\ c_i^y \\ c_i^z \end{pmatrix}$$

is called a non-uniform rational B-spline curve or *NURBS curve.*

A NURBS curve can be rewritten in order to define control points which possess a geometric meaning,

$$\mathbf{p}(t) = \sum_i R_i(t) \begin{pmatrix} p_i^x \\ p_i^y \\ p_i^z \end{pmatrix}$$

with the *weights* $w_i = c_i^w$, the *control points* $p_i^* = c_i^* / w_i$ and the *rational basis functions*

$$R_i(t) = \frac{1}{\sum_j w_j B_j(t)} w_i B_i(t).$$

The properties of the rational basis functions are summarized below.

Proposition 5.1. *The sum of the rational basis functions is equal to* 1. *If all weights are positive, then the rational basis functions are non-negative. The support of* R_j *is equal to the support of* B_j.

Similar to the case of B-splines, these facts imply certain geometric properties such as the convex hull property and affine invariance.

Proposition 5.2. *If all weights are positive, then a NURBS curve is contained in the convex hull of its control points. The description of a NURBS curve by its control points is invariant under affine transformations. If the boundary knots have multiplicity 1, then the first and last control points coincide with the first and last point of the curve.*

Increasing a weight causes the curve to move towards the corresponding control point. However, if all weights are increased simultaneously by the same factor, then the curve remains unchanged.

The evaluation of a NURBS curve should be performed by evaluating the three numerators and the denominator separately and then dividing the numerators by the denominator.

5.2 NURBS surfaces

A *bivariate tensor-product spline function* has the form

$$f(x,y) = \sum_{i,j} M_i(x) N_j(y) c_{ij}$$

where M_i and N_j are B-splines (possibly of different degree) with respect to two knot sequences T, U. Its domain is the Cartesian product of the two one-dimensional domains.

We consider four tensor-product spline functions $w(s,t), x(s,t), y(s,t)$ and $z(s,t)$ with knots $\mathcal{S}, \mathcal{T}$ and coefficients $c_{ij}^w, c_{ij}^x, c_{ij}^y, c_{ij}^z$. The piecewise rational surface

$$(s,t) \mapsto \mathbf{p}(s,t) = \frac{1}{w(s,t)} \begin{pmatrix} x(s,t) \\ y(s,t) \\ z(s,t) \end{pmatrix} =$$

$$= \frac{1}{\sum_{kl} M_k(s) N_l(t) c_{kl}^w} \sum_{ij} M_i(s) N_j(t) \begin{pmatrix} c_{ij}^x \\ c_{ij}^y \\ c_{ij}^z \end{pmatrix}$$

is called a non-uniform rational B-spline surface or *NURBS surface*.

Similar to the curve case, a NURBS surface can be rewritten in order to define control points which possess a geometric meaning.

$$\mathbf{p}(s,t) = \sum_{ij} R_{ij}(s,t) \begin{pmatrix} p^x_{ij} \\ p^y_{ij} \\ p^z_{ij} \end{pmatrix}$$

with the *weights* $w_{ij} = c^w_{ij}$, the *control points* $p^*_{ij} = c^*_{ij}/w_{ij}$ and the *rational basis functions*

$$R_{ij}(s,t) = \frac{1}{\sum_{kl} w_{kl} M_k(s) N_l(t)} w_{ij} M_i(s) N_j(t).$$

The properties of the rational basis functions are similar to the curve case. These facts imply again certain geometric properties (convex hull property, affine invariance), which are similar to the curve case.

The parameter lines of a NURBS surface are NURBS curves. If the multiplicity of the boundary knots of the knot vectors is equal to the degree, then the four boundary curves are the NURBS curves which are determined by the boundary control points and weights of the control mesh.

5.3 NURBS volumes

A *trivariate tensor-product spline function* has the form

$$f(x,y) = \sum_{i,j,k} L_i(x) M_j(y) N_k(z) c_{ijk}$$

where L_i, M_j and N_k are B-splines with respect to three knot sequences $\mathcal{S}, \mathcal{T}, \mathcal{U}$ (possibly of different degree). Its domain is the Cartesian product of the three one-dimensional domains.

We consider four tensor-product spline functions

$$w(r,s,t), x(r,s,t), y(r,s,t), z(r,s,t)$$

with knots S, T, U and coefficients $c^w_{ijk}, c^x_{ijk}, c^x_{ijk}, c^z_{ijk}$. The piecewise rational volume parameterization

$$\begin{aligned}(r,s,t) \mapsto \mathbf{p}(r,s,t) &= \frac{1}{w(r,s,t)} \begin{pmatrix} x(r,s,t) \\ y(r,s,t) \\ z(r,s,t) \end{pmatrix} \\ &= \frac{1}{\sum_{klm} L_k(r) M_l(s) N_m(t) c^w_{klm}} \sum_{hij} L_h(r) M_i(s) N_j(t) \begin{pmatrix} c^x_{hij} \\ c^y_{hij} \\ c^z_{hij} \end{pmatrix}\end{aligned}$$

is called a non-uniform rational B-spline volume or *NURBS volume.*

A NURBS volume can be rewritten in order to define control points which possess a geometric meaning,

$$\mathbf{p}(r,s,t) = \sum_{ijk} R_{ijk}(r,s,t) \begin{pmatrix} p^x_{ijk} \\ p^y_{ijk} \\ p^z_{ijk} \end{pmatrix}$$

with the *weights* $w_{ijk} = c^w_{ijk}$, the *control points* $p^*_{ijk} = c^*_{ijk}/w_{ijk}$ and the *rational basis functions*

$$R_{ijk}(r,s,t) = \frac{1}{\sum_{lmn} w_{lmn} L_l(r) M_m(s) N_n(t)} w_{ijk} L_i(r) M_j(s) N_k(t).$$

The properties of the rational basis functions are similar to the curve and surface case. These facts imply again certain geometric properties (convex hull property, affine invariance), which are similar to the curve and surface case.

The parameter lines of a NURBS volume are NURBS curves. If the multiplicity of the boundary knots of the knot vectors is equal to the degrees, then the 12 boundary curves are the NURBS curves which are determined by the boundary control points and weights of the control polygon.

The parameter surfaces (one parameter is kept constant) of a NURBS volume are NURBS surfaces. If the multiplicity of the boundary knots of the knot vectors is equal to the degrees, then the 6 boundary surfaces are the NURBS surfaces which are determined by the boundary control points and weights of the control mesh.

6 Simple isogeometric meshing techniques

We present a technique for generating NURBS parameterizations of circles and discuss the construction of NURBS parameterizations of planar quadrangular and and spatial hexahedral domains.

6.1 Parameterization of circles and spheres via inversion

A mapping is said to be *birational*, if both the mapping and its inverse are rational mappings, i.e., if they are both described by rational functions of the coordinates.

The *inversion at the unit sphere* is the birational mapping

$$\pi: \quad \mathbb{R}^d \cup \{\infty\} \to \mathbb{R}^d \cup \{\infty\}: \quad \mathbf{p} \mapsto \frac{1}{\mathbf{p}\cdot\mathbf{p}}\mathbf{p}.$$

Its domain is the d-dimensional space, which is completed by adding a single point at infinity. The point at infinity is mapped to the origin, and vice versa.

The inversion at the unit sphere satisfies $\pi^{-1} = \pi$. It maps the set of planes and sphere ($d = 3$) or lines and circles ($d = 2$) onto itself.

The inversion at the sphere can be used in order to generate NURBS representations of circles and spheres. For instance, applying it to the NURBS curve of degree 1,

$$w(t) = 1, \quad x(t) = 1, \quad y(t) = t, \quad t \in [-1, 1],$$

with knot vector $\mathcal{T} = (-1, 1)$, gives an arc of the circle with radius $\frac{1}{2}$ and center $(\frac{1}{2}, 0)$, with the two boundary points $(\frac{1}{5}, \pm\frac{2}{5})$. The control points and weights can be obtained with the help of the product formula for the blossoms or by inspection.

6.2 Gordon-Coons surfaces

Given four boundary curves of a quadrangular domain Ω in the plane, we want to find a NURBS parameterization of the domain,

$$\mathbf{p} : [0, 1]^2 \to \Omega.$$

This parameterization will be called a "isogeometric meshing" of the domain. It can be obtained using the construction of Gordon-Coons surfaces, as follows.

We define the two operators

$$\begin{array}{lll} A: & \mathcal{C}([0,1]^2, \mathbb{R}^d) \to \mathcal{C}([0,1]^2, \mathbb{R}^d): & \mathbf{p} \mapsto A\mathbf{p} \\ B: & \mathcal{C}([0,1]^2, \mathbb{R}^d) \to \mathcal{C}([0,1]^2, \mathbb{R}^d): & \mathbf{p} \mapsto B\mathbf{p} \end{array}$$

where

$$\begin{array}{l} (A\mathbf{p})(u, v) = (1 - u)\mathbf{p}(0, v) + u\mathbf{p}(1, v) \\ (B\mathbf{p})(u, v) = (1 - v)\mathbf{p}(u, 0) + v\mathbf{p}(u, 1). \end{array}$$

These two operators are projectors ($A^2 = A$, $B^2 = B$) and they commute ($AB = BA$). The parameterization

$$(u, v) \mapsto [(A + B - AB)\mathbf{p}](u, v)$$

is called the *Gordon-Coons parameterization* of the domain $\mathbf{p}([0, 1]^2)$.

The Gordon-Coons parameterization is fully determined by the four boundary curves. Consequently, it can be computed even if $\mathbf{p}$ is not known, provided that the (NURBS) parameterizations of the four boundary curves are available.

Theorem 6.1. *The Gordon-Coons parameterization matches the four boundary curves of the given patch* $\mathbf{p}$.

In order to prove this result, we observe that applying the operator A to the Gordon-Coons parameterization gives $A\mathbf{p}$, and applying B to it gives $B\mathbf{p}$.

- It should be noted that the Gordon-Coons parameterization is neither guaranteed to be regular nor is it guaranteed to be confined to the interior of the quadrangular region. Nevertheless, it gives reasonable results in many cases.
- In the case of NURBS boundary curves, the construction should be applied to the homogeneous coordinates

$$\mathbf{p}^h(u,v) = (w(u,v), x(u,v), y(u,v)),$$

  instead of applying it to the Cartesian coordinates. Otherwise the degrees and knot multiplicity would get very high.
- If opposite boundary knots have the same degrees and knot vectors, then these degrees and knot vectors are inherited by the Gordon-Coons NURBS patch.
- The construction can be extended to C^1 boundary conditions, by using Hermite interpolation of C^1 boundary data. In this case, however, the mixed first derivative vectors at the four corners of the patch have to be chosen in a consistent way. These vectors are called the "twist" vectors.
- The construction can also be used if some of the four given boundary curves degenerate into points. It is therefore possible to parametrize triangular and biangular domains, too. Domains with more than four boundaries can be dealt with by using non-differentiable boundary curves.

6.3 Gordon-Coons volumes

We outline the extension to the trivariate case.

Given six boundary patches of a hexahedral domain Ω in the plane, we want to find a NURBS parameterization of the domain,

$$\mathbf{p} : [0,1]^3 \to \Omega.$$

This parameterization will be called a "isogeometric meshing" of the domain.

We define the three operators

$$\begin{array}{llr} A: & \mathcal{C}([0,1]^3,\mathbb{R}^d) \to \mathcal{C}([0,1]^3,\mathbb{R}^d): & \mathbf{p} \mapsto A\mathbf{p}, \\ B: & \mathcal{C}([0,1]^3,\mathbb{R}^d) \to \mathcal{C}([0,1]^3,\mathbb{R}^d): & \mathbf{p} \mapsto B\mathbf{p}, \\ C: & \mathcal{C}([0,1]^3,\mathbb{R}^d) \to \mathcal{C}([0,1]^3,\mathbb{R}^d): & \mathbf{p} \mapsto C\mathbf{p} \end{array}$$

where

$$\begin{aligned} (A\mathbf{p})(u,v,w) &= (1-u)\mathbf{p}(0,v,w) + u\mathbf{p}(1,v,w) \\ (B\mathbf{p})(u,v,w) &= (1-v)\mathbf{p}(u,0,w) + v\mathbf{p}(u,1,w) \\ (C\mathbf{p})(u,v,w) &= (1-w)\mathbf{p}(u,v,0) + w\mathbf{p}(u,v,1) \end{aligned}$$

The three operators are projectors ($A^2 = A$, $B^2 = B$, $C^2 = C$) and they commute ($AB = BA$, $AC = CA$,$BC = CB$). The parameterization

$$(u,v) \mapsto [(A + B + C - AB - BC - CA + ABC)\mathbf{p}](u,v)$$

is called the *Gordon-Coons parameterization* of the domain $\mathbf{p}([0,1]^3)$.

The Gordon-Coons parameterization is then fully determined by the six boundary patches. Consequently, it can be computed even if $\mathbf{p}$ is not known, provided that the (NURBS) parameterizations of the six boundary patches are available.

Theorem 6.2. *The Gordon-Coons parameterization matches the six boundary patches of the given patch* $\mathbf{p}$.

In order to prove this result, we observe that applying the operator X to the Gordon-Coons parameterization gives $X\mathbf{p}$, where X is either A, B or C. The remarks about planar Gordon-Coons parameterization apply to volume parameterizations, too.

7 Further Reading

Among the rich literature about splines and NURBS, we particularly recommend the textbooks (Fa02; HL96; PBP02), the NURBS reference (PT95) and the collection of survey articles (FHK02) covering various aspects of the field of Computer Aided Geometric Design.

Bibliography

[Fa02] G. Farin: Curves and Surfaces for Computer Aided Geometric Design: A Practical Guide, Morgan–Kaufmann, 2002.

[FHK02] G. Farin, J. Hoschek, M.-S. Kim (eds.): Handbook of Computer Aided Geometric Design, Elsevier, 2002.

[HL96] J. Hoschek and D. Lasser: Fundamentals of Computer Aided Geometric Design, AK Peters, Wellesley, Mass., 1996.
[PT95] L. Piegl and W. Tiller: The NURBS book. Springer, Berlin 1995.
[PBP02] H. Prautzsch, W. Boehm and M. Paluszny: Bézier and B-Spline techniques. Springer, Berlin 2002.
[Ra89] Lyle Ramshaw: Blossoms are polar forms. Computer Aided Geometric Design 6(4): 323-358 (1989).

Extended Isogeometric Analysis for Strong and Weak Discontinuities

Vinh Phu Nguyen * and Stéphane Bordas †‡

* School of Civil, Environmental and Mining Engineering, The University of Adelaide, Adelaide, SA 5005, Australia

† School of Engineering, Institute of Mechanics and Advanced Materials, Cardiff University, Queen's Buildings, The Parade, Cardiff CF24 3AA, Wales, UK

‡ Faculté des Sciences, de la Technologie et de la Communication, University of Luxembourg, 6, rue Richard Coudenhove-Kalergi, 1359 Luxembourg City, Luxembourg

Abstract Isogeometric analysis (IGA) is a fundamental step forward in computational mechanics that offers the possibility of integrating methods for analysis into Computer Aided Design (CAD) tools and vice versa. The benefits of such an approach are evident, since the time taken from design to analysis is greatly reduced leading to large savings in cost and time for industry. The tight coupling of CAD and analysis within IGA requires knowledge from both fields and it is one of the goals of the present paper to outline much of the commonly used notation. In this manuscript, through a clear and simple Matlab® implementation, we present an introduction to IGA applied to the Finite Element (FE) method and related computer implementation aspects. Furthermore, implementation of the extended IGA which incorporates enrichment functions through the partition of unity method (PUM) is also presented, where several fracture examples in both two-dimensions and three-dimensions are given as illustration. The open source Matlab® code which accompanies the present paper can be applied to one, two and three-dimensional problems for linear elasticity, linear elastic fracture mechanics, structural mechanics (beams/plates/shells including large displacements and rotations) and Poisson problems with or without enrichment. It also implements the Bézier extraction concept that allows FE analysis to be performed on T-spline geometries. An attempt was also made to include a review of recent developments of IGA.

G. Beer, S.Bordas (Eds.), *Isogeometric Methods for Numerical Simulation*,
CISM International Centre for Mechanical Sciences
DOI 10.1007/ 978-3-7091-1843-6_2 © CISM Udine 2015

1 Introduction

The predominant technology that is used by CAD to represent complex geometries is the Non-Uniform Rational B-spline. This allows certain geometries to be represented exactly that can only be approximated by polynomial functions, such as conic and circular sections. There is a vast array of literature focused on Non-Uniform Rational B-splines (NURBS) (e.g. Piegl and Tiller (1996); Rogers (2001)) and as a result of several decades of research, many efficient computer algorithms exist for their fast evaluation and refinement. The key observation that was made by Hughes et al. Hughes et al. (2005) was that NURBS can not only be used to describe the geometry of a problem, but can also be used to construct finite approximations for analysis and coined the phrase 'Isogeometric Analysis' (IGA). Since this seminal paper, a monograph has been published entirely on the subject Cottrell et al. (2009) and applications have been found in several fields including structural mechanics, solid mechanics, fluid mechanics, fracture mechanics and contact mechanics.

In standard contact formulations using approximate geometry discretisations, faceted surfaces are often found on contact surfaces that can lead to jumps and oscillations in traction responses unless very fine meshes are used. The benefits of using NURBS over such an approach are therefore evident, since the contact surface is now smooth leading to more physically accurate contact stresses. Recent work in this area includes De Lorenzis et al. (2011); Jia (2011); Matzen et al. (2013); Temizer et al. (2011, 2012).

Another area where IGA has shown advantages over traditional approaches is optimisation problems Manh et al. (2011); Qian (2010); Qian and Sigmund (2011); Wall et al. (2008). IGA is particularly suited to such problems due to the tight coupling with CAD models and offers an extremely attractive approach for industrial applications. Moreover, it was found that certain numerical methods can be closely integrated with CAD where only a surface or boundary representation is required. In Scott et al. (2013); Simpson et al. (2012) an isogeometric boundary element method was presented where no meshing is required at all allowing CAD models to be used directly for analysis.

Shell and plate problems are another field of which IGA has demonstrated beneficial properties over conventional approaches Beirão da Veiga et al. (2012); Benson et al. (2010b, 2011, 2013); Echter et al. (2013); Kiendl et al. (2009); Uhm and Youn (2009). The smoothness of the NURBS basis functions allows for a straightforward construction of plate/shell elements. Particularly for thin shells, rotation-free formulations can be easily constructed Kiendl et al. (2009, 2010). Note that for multi-patch NURBS surfaces, rotation-free IGA elements require special treatment at patch boundaries where the basis functions are only C^0 continuous. In Kiendl et al. (2010), the bending strip method–a kind of penalty method was proposed to solve this problem. Furthermore, isogeometric plate/shell elements exhibit shear-locking much less pronouncedly compared to standard FE plate/shell elements. Elements with smooth boundaries such as circular and cylindrical elements were successfully constructed using the IGA concept J. (2009); Lu and Zhou (2011).

The smoothness of NURBS basis functions is attractive for analysis of fluids Bazilevs and Akkerman (2010); Gomez et al. (2010); Nielsen et al. (2011) and for fluid-structure interaction problems Bazilevs et al. (2008, 2009). In addition, due to the ease of constructing high order continuous basis functions, IGA has been used with great success in solving PDEs that incorporate fourth order (or higher) derivatives of the field variable such as the Hill-Cahnard equation Gómez et al. (2008), explicit gradient damage models Verhoosel et al. (2011b) and gradient elasticity Fischer et al. (2010). The high order NURBS basis have also found potential application in the Kohn-Sham equation for electronic structure modeling of semiconducting materials Masud and Kannan (2012).

IGA has been applied to cohesive fracture Verhoosel et al. (2011a) with a framework for modeling debonding material interfaces using NURBS and propagating cohesive cracks using T-splines. The method hinges on the ability to specify the continuity of NURBS/T-splines through a process known as knot insertion. As an extension of the eXtended Finite Element Method (XFEM) Moës et al. (1999), IGA has been applied to fracture using the partition of unity method (PUM) to capture two dimensional strong discontinuities and crack tip singularities efficiently De Luycker et al. (2011); Ghorashi et al. (2012). In Tambat and Subbarayan (2012) an explicit isogeometric enrichment technique was proposed for modeling material interfaces and cracks exactly. Note that this method is contrary to PUM-based enrichment methods which define the cracks implicitly. A phase field model for dynamic fracture was presented in Borden et al. (2012) where adaptive refinement with T-splines provides an effective method for simulating fracture in three dimensions. In Nguyen and Nguyen-Xuan (2013); Nguyen et al. (2014a) high order B-splines/NURBS were utilized to efficiently model delamination of two and three dimensional composite specimens.

NURBS are ubiquitous in CAD but were shown to exhibit major shortcomings, such as the inability to produce watertight geometries with no gaps and overlaps which often complicate mesh generation. From the perspective of analysis however, the major shortcoming is perhaps their tensor product structure of NURBS, where refinement must be a global operation. Refinement algorithms required for error estimation and adaptivity are therefore inefficient. As a solution to this, T-splines Sederberg et al. (2003) were recently developed to overcome the limitations of NURBS while making use of existing NURBS algorithms. T-splines correct the deficiencies of NURBS since local refinement and coarsening is permitted, and gap/overlaps are precluded. However, Buffa et al. (2010) note that linear independence of the T-spline basis functions is not guaranteed on generic T-meshes. Therefore, Scott et al. (2012) propose analysis-suitable T-splines, a mildly restricted subset of T-splines which appears to meet the demands of both design and analysis. Utilization of T-splines in an IGA framework has been illustrated in Bazilevs et al. (2010); Dörfel et al. (2010). Scott et al.Scott et al. (2011) showed that through an efficient Bézier extraction process, T-splines can be incorporated efficiently into existing FE codes.

Alternatives to T-splines are polycube splines Wang et al. (2008), PHT-splines (polynomial spline over hierarchical T-meshes) Deng et al. (2008) and LR-spline Dokken and Skytt (2010) among others. The PHT-splines were extended to rational splines for elastic continua and thin structures Nguyen-Thanh et al. (2011a,b). Adaptive refinement with PHT-splines is particularly simple. Though T-splines allow for local adaptive refinement, the complexity of knot insertion under adaptive refinement is complex, particularly in 3D. Yet another possibility is provided by hierarchical B-splines Bornemann and Cirak (2013); Schillinger et al. (2012); Vuong et al. (2011) and the unstructured Powell-Sabin splines Speleers et al. (2012). In particular, the hierarchical B-splines finite cell method Schillinger et al. (2012) furnishes a seamless CAD-FEA integration for very complex geometries. We refer also to Kleiss et al. (2012) for a IGA with finite element based local refinement capability. Different subdivision surface techniques (Catmull-Clark, Loop) were utilized for solids ans shells modeling Burkhart et al. (2010); Wawrzinek et al. (2011).

In computer aided geometric design, patching multiple NURBS to form complex topology is not easy for certain continuity between adjacent patches should be maintained. Trimming techniques provide a promising alternative for representing complex NURBS domains. In Kim et al. (2010) a trimmed surface based analysis framework has been proposed where the NURBS-enhanced FEM Sevilla et al. (2008, 2011) was applied to define a suitable integration domain within the parameter space. In a recent contribution Schmidt et al. (2012), the authors presented another method to deal with trimmed NURBS geometries. Discontinuous Galerkin or Nitsche's method was proposed to couple incompatible NURBS patches in Nguyen et al. (2014c); Ruess et al. (2013).

Some major computational aspects of IGA which have been studied so far include (i) locking issue, (ii) sensitivity to mesh distortion, (iii) impact of high continuity of NURBS on direct solvers, (iv) efficiency in large scale computations and (v) construction of a trivariate solid for a given bivariate surface representation.

Although the ultra-smoothness of NURBS basis functions reduces to some extent the locking phenomena for constrained problems e.g., incompressible media, thin-walled structures, NURBS-based FEs are not locking free Auricchio et al. (2007); Cardoso and Cesar de Sa (2012); Echter and Bischoff (2010); Elguedj et al. (2008a,b). Existing locking removal techniques in standard FEM were successfully adapted to IGA–the Discrete Strain Gap method Echter and Bischoff (2010), the F/B-bar method Elguedj et al. (2008a,b), the enhanced assumed strain method Cardoso and Cesar de Sa (2012) and mixed formulations Taylor (2011).

The effect of mesh distortion on the performance of IGA for solid mechanics was discussed in Lipton et al. (2010) in which it was found that higher-order NURBS functions are able to somewhat alleviate the impact of the distortions.

The high order continuity offered by NURBS has a negative impact on the performance of direct solvers as pointed out in Collier et al. (2012). The authors found that for a fixed number of unknowns and basis degree, a higher degree of continuity drastically increases the CPU time and RAM needed to solve the problem when using a direct solver.

In an attempt to compete with low order FEs with one-point quadrature that are extensively used in industrial applications, isogeometric collocation methods were developed Auricchio et al. (2010, 2012b).

Due to the fact that meshes in an isogeometric framework are defined by the parametrization of the object of interest, the quality of geometry parametrization plays an important role to assure mesh quality. This issue has, however, been addressed by only a few researchers, see e.g., Cohen et al. (2010); Schmidt et al. (2010); Takacs and Jüttler (2011); Xu et al. (2011); Zhou and Lu (2005). In particular, in Cohen et al. (2010), the authors proposed a concept of "analysis-aware geometry modeling".

In CAD, solids are actually surfaces for only the boundary not the interior is modeled. In FEA, a true solid is however needed. Therefore, the transition from CAD solids to FEA solids demands a step in which the CAD boundary representation will be converted to a FEA solid representation. Initial developments have been reported in Aigner et al. (2009); Escobar et al. (2011); Martin et al. (2009); Nguyen et al. (2014b); Zhang et al. (2007a, 2012a,b). Note that in this regard, the isogeometric boundary element method–IGABEM can be considered the truly isogeometric method, besides IGA plates/shells, up to date Peake et al. (2013); Scott et al. (2013);

Simpson et al. (2012, 2013).

Some implementation aspects of IGA have been reported in Cottrell et al. (2009) and more recently, an open source IGA Matlab® code has been described in Vuong et al. (2010) with a restriction to scalar PDEs. An additional open source IGA code written in Matlab® is given in de Falco et al. (2011) with applications including elasticity, scalar PDEs, magnetic problems etc. Incorporating IGA within an object-oriented C++ FE code is discussed in Rypl and Patzák (2012). Implementation details for enriched formulations within an IGA framework are reported in Benson et al. (2010a) using a commercial FE software.

In this chapter, an efficient easy-to-use IGA (X)FEM code for solid mechanics written in Matlab® (and some heavily used functions in C via MEX files) is presented. We believe that our code is very best suited for beginners (it follows the coding practice of the XFEM/EFG code described in Nguyen et al. (2008)) and it supports PUM-based enrichment for cracks and material interfaces. Various techniques (penalty method, Lagrange multipliers, least square method) for Dirichlet boundary conditions enforcement are provided. Both two and three dimensional elasticity are supported and a 2D/3D (in)compatible multi-patch IGAFEM is also implemented. Visualization can be performed directly in Matlab or in Paraview. The code supports T-splines surfaces that are created using Rhino3d and the T-splines plugin.

The remainder of the chapter is structured as follows: Section 2 presents the basis of B-splines and NURBS curves, surfaces and solids. Section 3 is devoted to the Bubnov-Galerkin IGA FEM with emphasis on the differences with standard FEM. The IGA (X)FEM code written in Matlab®, the MIGFEM, is treated in Section 4. An extended isogeometric formulation based on the concept of PUM is a subject of Section 5. Numerical examples on solid and fracture mechanics are given in Section 6. Section 7 discusses the isogeometric beam and shell formulations.

2 Brief recall on B-splines and NURBS

In this section a brief discussion about B-spline and its powerful generalization, NURBS is given. We begin with some basic concepts including knot vectors, B-spline basis functions. Then, B-spline curves and surfaces are introduced. We refer to Piegl and Tiller (1996) for more details.

2.1 Parametric representation

In order to present the basic concepts such as parameters, parametric space and physical space, let us consider the equation of a unit circle centered at the origin

$$x^2 + y^2 = 1 \tag{1}$$

which can be rewritten using the so-called parametric equation

$$x = \cos t, \quad y = \sin t \quad 0 \leq t \leq 2\pi \tag{2}$$

in which t denotes a parameter ranging from 0 to 2π. The interval $[0, 2\pi]$ constitutes a parametric space. Equation (2) maps a parameter (t) in the parametric space to a point (x, y) on the unit circle–the physical space.

2.2 Knot vectors

A *knot vector* (defined for a given direction) is a sequence (in ascending order) of parameter values, written $\Xi = \{\xi_1, \xi_2, \ldots, \xi_{n+p+1}\}$ where ξ_i is the ith knot, n is the number of basis functions, p is the polynomial order. The knot vector divides the parametric space into intervals usually referred to as *knot spans*. Consecutive knots can have the same value i.e., more than one knot can be located at the same coordinate in the parametric space. A number of coinciding knots is referred to as a knot with a certain *multiplicity*. A knot vector is said to be *open* if its first and last knots having $p + 1$ multiplicity. An important property of open knot vectors is that basis functions formed by them are interpolatory at the ends of the parametric space. A *uniform knot vector* is associated to evenly distributed knots. Otherwise it is said to be a non-uniform knot vector.

2.3 Basis functions

Given a knot vector Ξ, the B-spline basis functions are defined recursively starting with the zeroth order basis function ($p = 0$) given by

$$N_{i,0}(\xi) = \begin{cases} 1 & \text{if } \xi_i \leq \xi \leq \xi_{i+1}, \\ 0 & \text{otherwise.} \end{cases} \tag{3}$$

and for $p \geq 1$

$$N_{i,p}(\xi) = \frac{\xi - \xi_i}{\xi_{i+p} - \xi_i} N_{i,p-1}(\xi) + \frac{\xi_{i+p+1} - \xi}{\xi_{i+p+1} - \xi_{i+1}} N_{i+1,p-1}(\xi) \tag{4}$$

This is referred to as the **Cox-de Boor recursion formula**. Note that when evaluating these functions, ratios of the form 0/0 are defines as zeros.

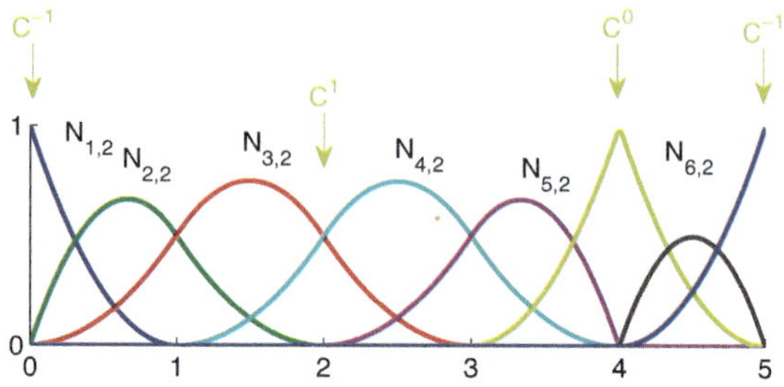

Figure 1: Quadratic B-spline basis functions for a open non-uniform knot vector $\Xi = \{0, 0, 0, 1, 2, 3, 4, 4, 5, 5, 5\}$. There are eight functions. Note that the basis functions at the beginning and end of the domain ($N_{1,2}$ and $N_{8,2}$) are not identical from the ones in the interior.

Some salient properties of B-spline basis functions are (1) they constitute a partition of unity, (2) each basis function is nonnegative over the entire domain, (3) they are linearly independent, (4) the support of a B-spline function of order p is $p+1$ knot spans i.e., $N_{i,p}$ is non-zero over $[\xi_i, \xi_{i+p+1}]$, (5) basis functions of order p have $p - m_i$ continuous derivatives across knot ξ_i where m_i is the multiplicity of knot ξ_i and (6) B-spline basis are generally only approximants and not interpolants. Therefore imposing Dirichlet boundary conditions will not be as straightforward as in FEM.

Figure 1 illustrates similar functions for an open non-uniform knot vector. Note that the basis functions are interpolatory at each end of the interval and also at $\xi = 4$, the location of a repeated knot, where only C^0-continuity is attained. Elsewhere, the functions are C^1-continuous (C^{p-1}). The ability to control continuity by means of knot insertion is particularly useful for modeling discontinuities such as interfacial cracks or material interfaces Nguyen et al. (2014a); Verhoosel et al. (2011a) although it requires knots to be placed on the discontinuity surfaces which is not as flexible as PUM enrichment.

2.4 B-spline curves

Let the number of space dimensions be denoted by d. Given n B-splines basis functions $N_{i,p}$ and n *control points* $\mathbf{B}_i \in \mathbb{R}^d$, a piecewise-polynomial B-spline curve is given by

$$\mathbf{C}(\xi) = \sum_{i=1}^{n} N_{i,p}(\xi)\mathbf{B}_i \tag{5}$$

in words, a B-spline curve is constructed as a linear combination of the B-spline basis functions with control points as the coefficients. Piecewise linear interpolation of the control points defines the *control polygon.*

2.5 B-spline surfaces

Given two knot vectors (one for each direction) $\Xi = \{\xi_1, \xi_2, \ldots, \xi_{n+p+1}\}$ and $\mathcal{H} = \{\eta_1, \eta_2, \ldots, \eta_{m+q+1}\}$ and a control net $\mathbf{B}_{i,j} \in \mathbb{R}^d$, a tensor-product B-spline surface is defined as

$$\mathbf{S}(\xi, \eta) = \sum_{i=1}^{n}\sum_{j=1}^{m} N_{i,p}(\xi)M_{j,q}(\eta)\mathbf{B}_{i,j} \tag{6}$$

where $N_{i,p}(\xi)$ and $M_{j,q}(\eta)$ are the univariate B-spline basis functions of order p and q corresponding to knot vectors Ξ and $\mathcal{H}$, respectively. The above can be rewritten in a more compact form as follows

$$\mathbf{S}(\xi, \eta) = \sum_{i=1}^{n}\sum_{j=1}^{m} N_{i,j}^{p,q}(\xi, \eta)\mathbf{B}_{i,j} \tag{7}$$

where $N_{i,j}^{p,q}(\xi, \eta) = N_{i,p}(\xi)M_{j,q}(\eta)$ are the bivariate B-spline basis function. The parametric domain is a rectangle $(\xi, \eta) \in [\xi_1, \xi_{n+p+1}] \times [\eta_1, \eta_{m+q+1}]$. In Figure 2 a bi-cubic B-spline basis function is presented.

2.6 NURBS

B-splines are convenient for free-form modelling, but they lack the ability to exactly represent some simple shapes like circles and ellipsoids. This is why today, the de facto standard technology in CAD is a generalization of B-splines called NURBS. NURBS stands for Non-Uniform Rational B-splines, they are rational functions of B-splines and inherit all their favourable properties. NURBS extend B-splines since they allow exact representation of conic sections. The major strengths of NURBS are that they are convenient for free-form surface modeling, can exactly represent all quadric surfaces, such as cylinders, spheres, ellipsoids, etc., and that there exist many efficient and numerically stable algorithms to generate NURBS objects Piegl and Tiller (1996).

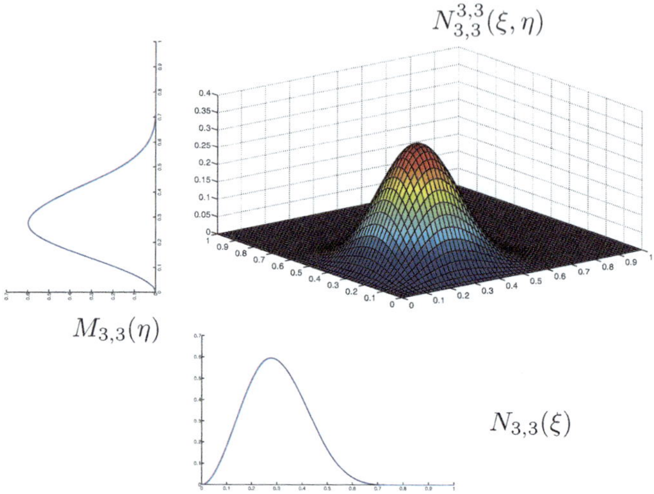

Figure 2: A bivariate cubic B-spline basis function.

NURBS are commonly used in computer-aided design (CAD), manufacturing (CAM), and engineering (CAE) and are part of numerous industry wide used standards, such as IGES, STEP, ACIS, and PHIGS.

NURBS basis functions A rational basis functions is defined as

$$R_{i,p}(\xi) = \frac{N_{i,p}(\xi)w_i}{W(\xi)} = \frac{N_{i,p}(\xi)w_i}{\sum_{j=1}^{n} N_{j,p}(\xi)w_j} \tag{8}$$

where w_i are a set of n positive weights. Selecting appropriate values for the w_i permits the description of many different types of curves including polynomials and circular arcs. For the special case in which $w_i = c, i = 1, 2, \ldots, n$ the NURBS basis reduces to the B-spline basis due to the partition of unity property of B-spline basis functions.

NURBS curves and surfaces The NURBS curve is defined, in the same manner as the B-spline curves, as

$$\mathbf{C}(\xi) = \sum_{i=1}^{n} R_{i,p}(\xi)\mathbf{B}_i \tag{9}$$

The NURBS surfaces are defined as

$$S(\xi,\eta) = \sum_{i=1}^{n}\sum_{j=1}^{m} R_{i,j}^{p,q}(\xi,\eta)\mathbf{B}_{i,j} \tag{10}$$

where $R_{i,j}$ are given by

$$R_{i,j}(\xi,\eta) = \frac{N_i(\xi)M_j(\eta)w_{i,j}}{\sum_{\hat{i}=1}^{n}\sum_{\hat{j}=1}^{m} N_{\hat{i}}(\xi)M_{\hat{j}}(\eta)w_{\hat{i},\hat{j}}} \tag{11}$$

3 Finite element analysis with NURBS

3.1 Isoparametric discretisation

The isoparametric concept refers to the use of the same basis functions for both geometry and unknown field discretisation. Both FEM and IGA formulations most commonly utilise the isoparametric concept. However, they differ from each other by the type of basis functions used. That is,

- Finite elements: the basis which is chosen to approximate the unknown field, is also used to approximate the known geometry. This most commonly takes the form of (low order) polynomial functions and the geometry is in most cases only approximated.
- Isogeometric analysis: the basis is chosen to exactly capture the geometry and this is also used to approximate the unknown field. Refinement may be required for the unknown fields, but the exact geometry is maintained at all stages of analysis.

The physical domain is denoted by Ω and the parametric domain by $\hat{\Omega}$, where we refer to Figure 3 for the illustration of these terms. In two dimensions, the parametric domain is a square and in three dimensions it is a cube. The mapping from the parametric domain to the physical domain is then given by

$$\mathbf{x} = \sum_{I=1}^{n} \Phi_I(\boldsymbol{\xi})\mathbf{B}_I, \tag{12}$$

where the shape function Φ_I refers to either the univariate NURBS basis function if Ω is a curve or the bivariate NURBS basis function in case Ω is a surface. The number of control points is denoted by n. In the above, $\boldsymbol{\xi}$ is the parametric coordinate i.e., ξ in one dimension, (ξ,η) in two dimensions and $\mathbf{B}_I$ denotes control point number I.

In an isoparametric formulation, the unknown field variable is approximated by the same shape functions as the geometry. That is,

$$\mathbf{u}(\mathbf{x}) = \sum_{I=1}^{n} \Phi_I(\boldsymbol{\xi})\mathbf{u}_I, \tag{13}$$

where $\mathbf{u}_I$ denotes the value of the field variable at control point $\mathbf{B}_I$. It is therefore referred to as a *control variable* or more generally a degree of freedom (dof). It should be emphasized that since control points do not generally locate inside the domain of interest or even fall outside the physical domain $\mathbf{u}_I$ does not usually represent a physical nodal value as in conventional FEM. For control points inside the physical domain, $\mathbf{u}_I$ also does not represent a physical nodal displacement due to the non-interpolatory nature of NURBS. This is similar to the case of meshless methods built on non-interpolatory shape functions such as the moving least squares (MLS) Belytschko et al. (1994a); Nayroles et al. (1992); Nguyen et al. (2008).

3.2 Spatial derivatives of shape functions

Before computing the spatial derivatives $N_{I,x}$ and $N_{I,y}$ of the shape functions, it is necessary to compute the Jacobian matrix of the geometry mapping defined in Equation (12)

$$\mathbf{J}_{\boldsymbol{\xi}} = \begin{bmatrix} x_{,\xi} & x_{,\eta} \\ y_{,\xi} & y_{,\eta} \end{bmatrix}, \tag{14}$$

with the components calculated as (using Equation (12))

$$\frac{\partial x_i}{\partial \xi_j} = \frac{\partial \Phi_I}{\partial \xi_j} B_{iI}, \tag{15}$$

where B_{iI} represents the ith coordinate of control point I. The determinant of $\mathbf{J}_{\boldsymbol{\xi}}$ is denoted by $|J_{\boldsymbol{\xi}}|$.

The derivatives of basis functions with respect to the physical domain coordinates are given by

$$\begin{bmatrix} \Phi_{I,x} & \Phi_{I,y} \end{bmatrix} = \begin{bmatrix} \Phi_{I,\xi} & \Phi_{I,\eta} \end{bmatrix} \begin{bmatrix} \xi_{,x} & \xi_{,y} \\ \eta_{,x} & \eta_{,y} \end{bmatrix} = \begin{bmatrix} \Phi_{I,\xi} & \Phi_{I,\eta} \end{bmatrix} \mathbf{J}_{\boldsymbol{\xi}}^{-1}, \tag{16}$$

where $\Phi_{I,\xi}$ and $\Phi_{I,\eta}$ can be found in textbook Piegl and Tiller (1996).

3.3 Numerical integration

Integrals over the entire geometry (physical domain Ω) are split into elemental integrals over a domain denoted by Ω_e. These integrals are pulled back to the parametric element $\hat{\Omega}_e$ via the geometry mapping, Equation (12). Finally, integrals over the parametric element are pulled back to the parent domain $\square$ (the determinant of the Jacobian is $|J_{\bar{\boldsymbol{\xi}}}|$) where existing integration rules are usually defined. Refer to Figure 3 for an illustration. We write for a function f of two variables x and y

$$\begin{aligned}\int_{\Omega} f(x,y)\mathrm{d}\Omega &= \sum_{e=1}^{n}\int_{\Omega_e} f(x,y)\mathrm{d}\Omega_e = \sum_{e=1}^{n}\int_{\hat{\Omega}_e} f(x(\xi),y(\eta))|J_{\boldsymbol{\xi}}|\mathrm{d}\hat{\Omega}_e \\ &= \sum_{e=1}^{n}\int_{\square} f(\bar{\xi},\bar{\eta})|J_{\boldsymbol{\xi}}||J_{\bar{\boldsymbol{\xi}}}|\mathrm{d}\square.\end{aligned} \tag{17}$$

The final integral can be computed using standard Gauss-Legendre quadrature (hereafter named Gaussian). Specially, a $(p+1)\times(q+1)$ Gaussian quadrature is adopted for two dimensional elements with p and q denoting the orders of the NURBS basis in the ξ and η directions. It should be emphasized that Gaussian quadrature is not optimal for IGA. In Auricchio et al. (2012a); Hughes et al. (2010), an optimal quadrature rule–the half-point rule–for IGA was presented. In our paper, only a standard Gauss-Legendre quadrature implementation is used.

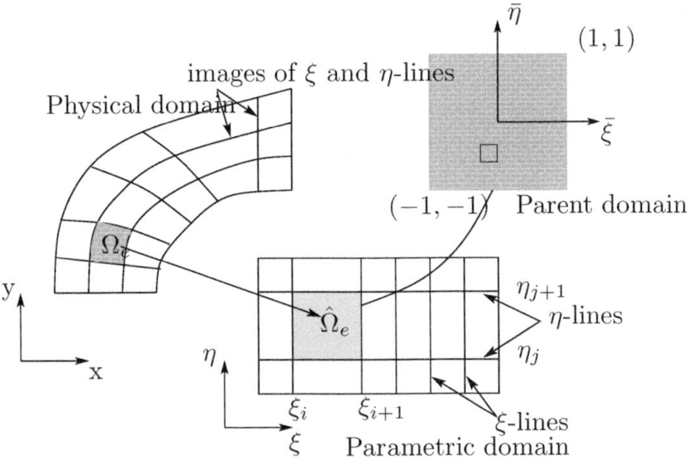

Figure 3: Definition of domains used for integration in isogeometric analysis. Elements are defined in the parametric space as non-zero knot spans, $[\xi_i, \xi_{i+1}] \times [\eta_j, \eta_{j+1}]$ and elements in the physical space are images of their parametric counterparts.

The transformation from the parent domain □ to a parametric domain $[\xi_i, \xi_{i+1}] \times [\eta_i, \eta_{i+1}]$ is given by

$$\xi = \frac{1}{2}[(\xi_{i+1} - \xi_i)\bar{\xi} + (\xi_{i+1} + \xi_i)] \tag{18}$$

$$\eta = \frac{1}{2}[(\eta_{j+1} - \eta_j)\bar{\eta} + (\eta_{j+1} + \eta_j)]. \tag{19}$$

where $\bar{xi}, \bar{\eta}$ represents a Gauss point. Therefore, the determinant of the Jacobian of this transformation reads

$$|J_{\bar{\boldsymbol{\xi}}}| = \frac{1}{4}(\xi_{i+1} - \xi_i)(\eta_{j+1} - \eta_j). \tag{20}$$

3.4 Assembly for one dimensional problems

To illustrate the use of the isogeometric concept in a finite element context, an example is best suited for explanation. Let us consider the following one dimensional Poisson equation

$$u_{,xx}(x) + b(x) = 0 \quad x \in [0,1]; \quad u(0) = 0, \quad u(1) = 0, \tag{21}$$

with $b(x) = x$. It should be emphasized that the boundary conditions given above are homogeneous Dirichlet boundary conditions which can be imposed directly as in FEM by setting the corresponding control variables as zeros. This is due to the use of an open knot vector which ensures that the first and last points of the curve are interpolatory (see Section 2.6). More details on imposition of boundary conditions are provided in Section 3.6. The exact solution for this problem is

$$u(x) = -\frac{1}{6}x^3 + \frac{1}{6}x. \tag{22}$$

Using the method of weighted residuals, the weak form associated with Equation (21) writes

$$\text{Find } u \in \mathcal{U} \mid \forall v \in \mathcal{U}^o \ , \quad \int_0^1 \frac{du}{dx}\frac{dv}{dx}\mathrm{d}x = \int_0^1 b(x)v(x)\mathrm{d}x, \tag{23}$$

where $\mathcal{U}$ and $\mathcal{U}^o$ are the usual trial and test spaces. Using n shape functions N_I as test and trial functions, the discrete form of this equation reads

$$\text{Find } (u_J)_{1\leq j\leq n} \in \mathbb{R}^n \ , \quad \left(\int_0^1 N_{I,x}(x)N_{J,x}(x)\mathrm{d}x\right) u_J = \int_0^1 b(x)N_I(x)\mathrm{d}x, \tag{24}$$

using Equation (13). This can be written in matrix form

$$K_{IJ} u_J = f_I, \tag{25}$$

where the stiffness matrix and the force vector are defined as

$$\forall I, J \in \{1, \ldots, n\}, \quad K_{IJ} = \int_0^1 N_{I,x}(x) N_{J,x}(x) \mathrm{d}\Omega, \quad f_I = \int_0^1 b(x) N_I(x) \mathrm{d}\Omega. \tag{26}$$

In order to build the global stiffness matrix, first the elementary stiffness matrices are computed. They are then assembled into the global stiffness matrix $\mathbf{K}$. In the same manner, one obtains the global force vector $\mathbf{f}$. The solution is then given by $\mathbf{u} = \mathbf{K}^{-1}\mathbf{f}$.

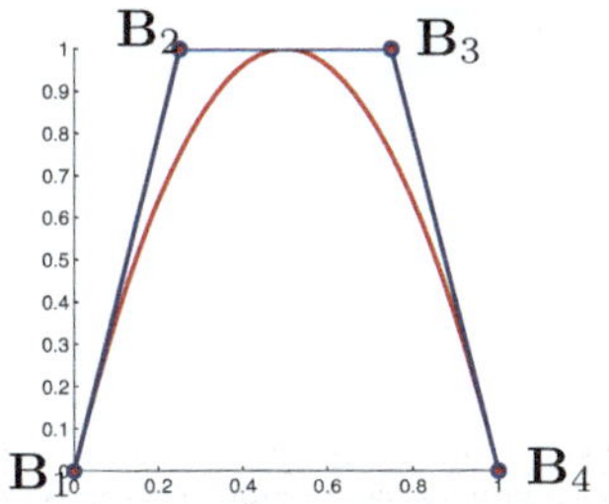

A quadratic B-spline curve with 4 control points

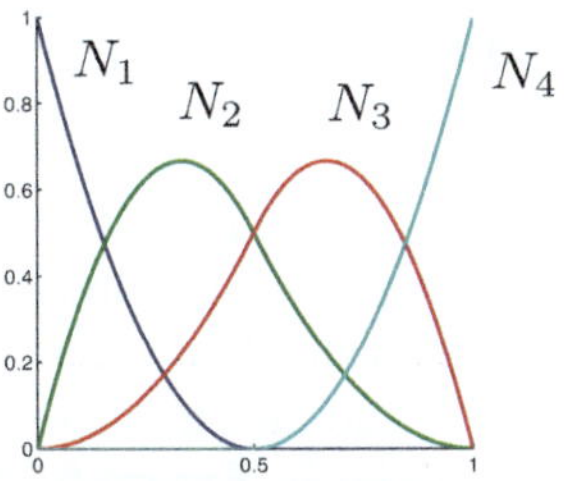

4 quadratic basis functions

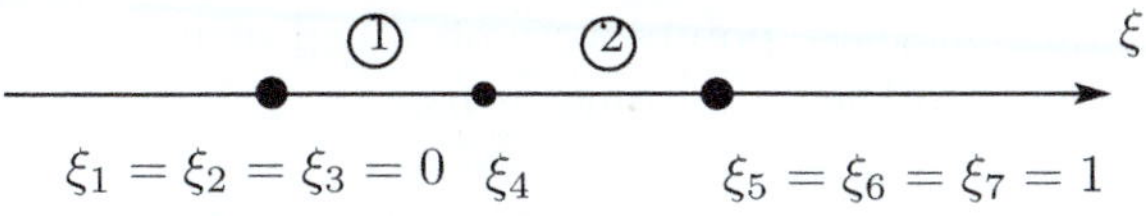

Parametric space

Figure 4: One dimensional isogeometric analysis example: exact geometry, quadratic basis functions and mesh in the parametric space.

To illustrate the assembly of the elementary stiffness into the global stiffness matrix, let us consider the example given in Figure 4. Recall that for element $e = [\xi_i, \xi_{i+1}]$, there are $p+1$ non-zero basis functions N_α with $i - p \leq \alpha \leq i$. This can be seen from Figure 4. For this problem, the data is therefore as follows

$$\begin{array}{ccccc} \text{element} & \text{range} & \text{non-zero basis} & \text{dofs} & \text{control points} \\ 1 & [\xi_3, \xi_4] & N_1, N_2, N_3 & u_1, u_2, u_3 & \mathbf{B}_1, \mathbf{B}_2, \mathbf{B}_3 \\ 2 & [\xi_4, \xi_5] & N_2, N_3, N_4 & u_2, u_3, u_4 & \mathbf{B}_2, \mathbf{B}_3, \mathbf{B}_4 \end{array} , \qquad (27)$$

where the dof data is used to assemble the global stiffness matrix. One can consider this mesh as composed of two elements of which each has three nodes (control points). In this manuscript we use nodes and control points interchangeably.

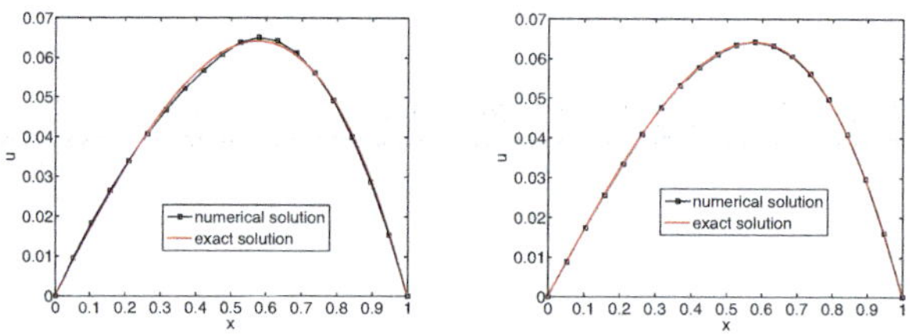

(a) 2 elements (b) 4 elements

Figure 5: Comparison of the IGA result against the exact solution for the one dimensional PDE given in Equation (21).

The good agreement of the numerical results for quadratic B-spline elements compared to the exact solution as shown in Figure 5 verifies the implementation. For the coarse mesh of two quadratic elements, the knot vector is $\Xi = [0, 0, 0, 0.5, 1, 1, 1]$ and hence there are 4 CPs and 4 nodal unknowns. For smooth curves given in Figure 5, 20 uniformly distributed points on the parameter space (the interval $[0, 1]$) were used. The coordinates and u values at these points were obtained using interpolations from the control points and the nodal unknowns.

3.5 Assembly process for two dimensional elastostatic analysis

Consider a domain Ω, bounded by Γ. The boundary is partitioned into two sets: Γ_u and Γ_t with displacements prescribed on Γ_u and tractions $\bar{\mathbf{t}}$ prescribed on Γ_t. The weak form of a linear elastostatics problem is to find

$\mathbf{u}$ in the trial space [1], such that for all test functions $\delta\mathbf{u}$ in the test space [2],

$$\int_{\Omega} \varepsilon(\mathbf{u}) : \mathbf{D} : \varepsilon(\delta\mathbf{u}) \mathrm{d}\Omega = \int_{\Gamma_t} \bar{\mathbf{t}} \cdot \delta\mathbf{u} \mathrm{d}\Gamma + \int_{\Omega} \mathbf{b} \cdot \delta\mathbf{u} \mathrm{d}\Omega, \tag{28}$$

where the elasticity matrix is denoted by $\mathbf{D}$ and $\mathbf{b}$ refers to a body force. Using the Bubnov-Galerkin method where the same shape functions Φ_I are used for both $\mathbf{u}$ and $\delta\mathbf{u}$, we can write

$$\mathbf{u}(\mathbf{x}) = \sum_{I}^{n} \Phi_I(\boldsymbol{\xi})\mathbf{u}_I, \quad \delta\mathbf{u}(\mathbf{x}) = \sum_{I}^{n} \Phi_I(\boldsymbol{\xi})\delta\mathbf{u}_I, \tag{29}$$

where $\delta\mathbf{u}_I$ denote the nodal displacement variations.

Substitution of these approximations into Equation (28) and using the arbitrariness of the nodal variations gives the discrete set of equations

$$\mathbf{K}\,\mathbf{u} = \mathbf{f}, \tag{30}$$

with

$$\mathbf{K}_{IJ} = \int_{\Omega} \mathbf{B}_I^{\mathrm{T}} \mathbf{D} \mathbf{B}_J \mathrm{d}\Omega, \quad \mathbf{f}_I = \int_{\Gamma_t} \Phi_I \bar{\mathbf{t}} \mathrm{d}\Gamma + \int_{\Omega} \Phi_I \mathbf{b} \mathrm{d}\Omega. \tag{31}$$

In two dimensions, the $\mathbf{B}$ matrix (not to be confused with the control net) is given by

$$\mathbf{B}_I = \begin{bmatrix} \Phi_{I,x} & 0 \\ 0 & \Phi_{I,y} \\ \Phi_{I,y} & \Phi_{I,x} \end{bmatrix}, \tag{32}$$

where the shape function derivatives are computed according to Equation (16).

We now consider the two dimensional problem shown in Figure 6. In this case, the knot vectors are $\Xi = [0, 0, 1, 1]$ and $\mathscr{H} = [0, 0, 0, 0.5, 1, 1, 1]$, respectively. The orders of the basis functions are $p = 1$ and $q = 2$. Eight control points are defined: two points along the ξ direction and four points along the η direction.

The domain and non-zero basis functions for Element 1 are given by

$$\begin{array}{ccc} \text{direction} & \text{range} & \text{non zero basis} \\ \xi & [\xi_2, \xi_3] & N_1, N_2 \\ \eta & [\eta_3, \eta_4] & M_1, M_2, M_3 \end{array} \tag{33}$$

[1] contains C^0 functions

[2] contains C^0 functions vanishing on Γ_u

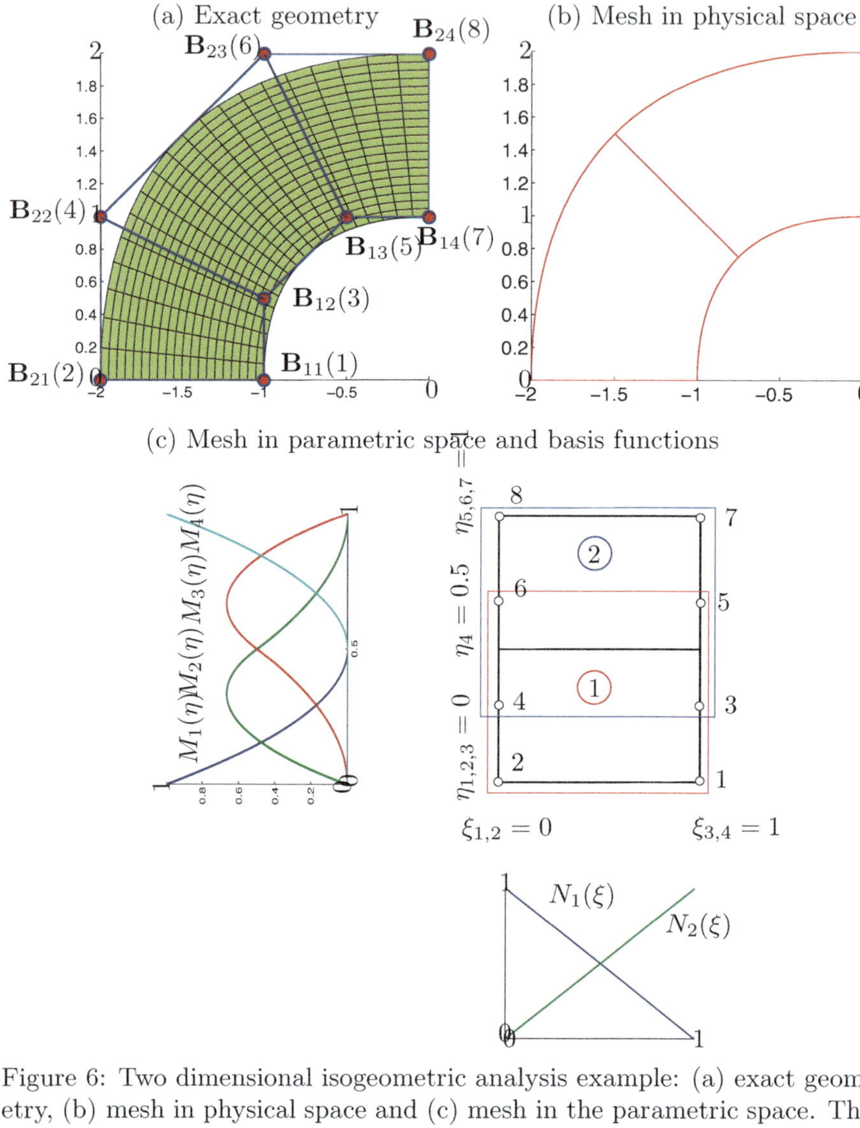

Figure 6: Two dimensional isogeometric analysis example: (a) exact geometry, (b) mesh in physical space and (c) mesh in the parametric space. The two rectangles are used to illustrate the control points belonging to each element.

Hence there are six non-zero basis functions on element $e = 1$ which can be assembled into a vector $\mathbf{R}$ as follows

$$\mathbf{R}_1 = [N_1M_1, N_2M_1, N_1M_2, N_2M_2, N_1M_3, N_2M_3]. \tag{34}$$

These six basis functions are associated with six global basis indices given by [3]

$$\text{element}(1,:) = [1,2,3,4,5,6]. \tag{35}$$

Similarly, for element 2, the shape function vector is given by

$$\mathbf{R}_2 = [N_1M_2, N_2M_2, N_1M_3, N_2M_3, N_1M_4, N_2M_4], \tag{36}$$

with the associated global indices

$$\text{element}(2,:) = [3,4,5,6,7,8]. \tag{37}$$

Control points are stored in a two dimensional matrix of dimensions $n \times 2$ (n denotes the number of control points in the mesh). The connectivity data is stored in a two dimensional matrix of dimensions $n_{el} \times (p+1)*(q+1)$ where n_{el} denotes the number of elements. For the example under consideration, these two matrices are given by

$$\text{controlPts} = \begin{bmatrix} \mathbf{B}_{11} \\ \mathbf{B}_{21} \\ \mathbf{B}_{12} \\ \mathbf{B}_{22} \\ \mathbf{B}_{13} \\ \mathbf{B}_{23} \\ \mathbf{B}_{14} \\ \mathbf{B}_{24} \end{bmatrix}, \quad \text{element} = \begin{bmatrix} 1 & 2 & 3 & 4 & 5 & 6 \\ 3 & 4 & 5 & 6 & 7 & 8 \end{bmatrix}. \tag{38}$$

The knot ranges along the ξ and η directions are stored in the following matrices

$$\text{elRangeU} = \begin{bmatrix} 0 & 1 \end{bmatrix}, \quad \text{elRangeV} = \begin{bmatrix} 0 & 0.5 \\ 0.5 & 1 \end{bmatrix}, \tag{39}$$

where the number of rows is equal to the number of elements (non-zero knot spans) in each direction.

With this vector of basis functions $\mathbf{R}$, we can compute the derivatives of the basis functions. The derivatives of the basis functions with respect to x are stored in the following vector

[3] in Matlab® notation.

$$\mathbf{R}_{,x} = \begin{bmatrix} R_{1,x} & R_{2,x} & R_{3,x} & R_{4,x} & R_{5,x} & R_{6,x} \end{bmatrix}^{\mathrm{T}}, \tag{40}$$

Similarly, we have $\mathbf{R}_{,y}$ for the derivatives of the basis functions with respect to y. Having these basis function derivatives, we are now ready to define the $\mathbf{B}$ matrix for any element e

$$\mathbf{B}_e = \begin{bmatrix} \mathbf{R}[1]_{,x} & \mathbf{R}[2]_{,x} & \cdots & \mathbf{R}[6]_{,x} & 0 & \cdots & 0 \\ 0 & 0 & \cdots & 0 & \mathbf{R}[1]_{,y} & \cdots & \mathbf{R}[6]_{,y} \\ \mathbf{R}[1]_{,y} & \mathbf{R}[2]_{,y} & \cdots & \mathbf{R}[6]_{,y} & \mathbf{R}[1]_{,x} & \cdots & \mathbf{R}[6]_{,x} \end{bmatrix}, \tag{41}$$

where the control point displacement vector is stored in the following order $\mathbf{u} = [u_{1x}, u_{2x}, \ldots, u_{nx}, u_{1y}, u_{2y}, \ldots, u_{ny}]^{\mathrm{T}}$. The element stiffness matrix is then given by

$$\mathbf{K}_e = \int_{\Omega_e} \mathbf{B}_e^{\mathrm{T}} \mathbf{D} \mathbf{B}_e \mathrm{d}\Omega_e, \tag{42}$$

which is then assembled to the global stiffness matrix using the element connectivity matrix.

In order to compute the external force vector, it is convenient to define a boundary mesh as shown in Figure 7. The computation of the external force vector $\int_\Gamma \mathbf{R}^{\mathrm{T}} \bar{\mathbf{t}} \mathrm{d}\Gamma$ then follows the one dimensional assembly procedure given in Section 3.4.

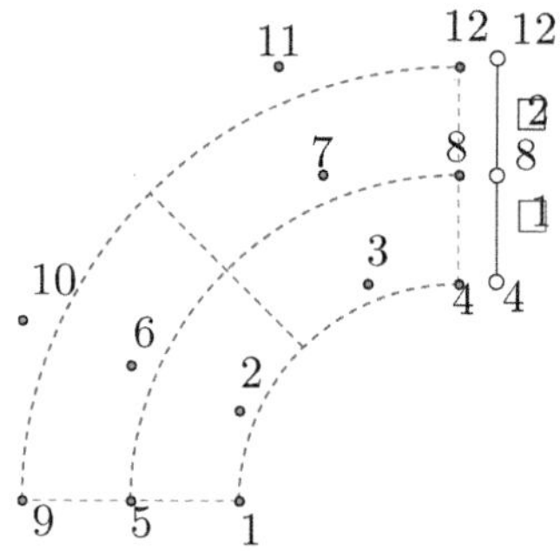

Figure 7: Boundary mesh for external force computation. A linear basis is used in the η direction. Assume a traction is applied on the edge containing nodes 4, 8 and 12. The boundary mesh is composed of two linear isogeometric elements 1 and 2.

3.6 Boundary condition enforcement

Figure 8 illustrates two kinds of Dirichlet boundary conditions: on edge AD, $u_x = 0$ and on edge BC, $u_y = \bar{u}$. The former is a homogeneous Dirichlet boundary condition (BC) while the latter is referred to as uniform inhomogeneous Dirichlet BCs. Homogeneous Dirichlet BCs can be enforced by setting the corresponding control variables as zeros (in this example, setting $u_{xI} = 0, I = 2, 4, 6, 8$). For edge BC, the inhomogeneous Dirichlet BCs can also be satisfied by setting $u_{yI} = \bar{u}, I = 1, 3, 5, 7$. This is due to the partition of unity property of the NURBS basis i.e., $u_y^{BC} = M_1(\eta)u_{y1} + M_2(\eta)u_{y3} + M_3(\eta)u_{y5} + M_4(\eta)u_{y7} = (M_1(\eta) + M_2(\eta) + M_3(\eta) + M_4(\eta))\bar{u} = \bar{u}$. Inhomogeneous Dirichlet BCs applied on corner control points (black points in Figure 8) are enforced by simply setting the corner control variables equal to the prescribed values since the NURBS shape functions at these points satisfy the Kronecker delta property (assuming the use of open knot vectors). This is called direct imposition of Dirichlet BCs.

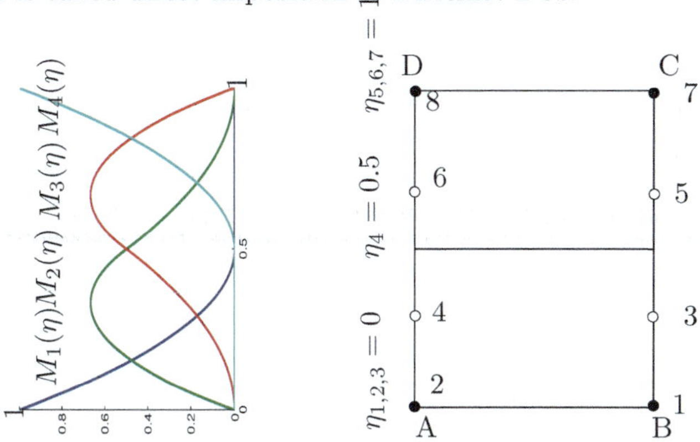

Figure 8: Imposing Dirichlet BCs: black points denote corner control points where the NURBS basis satisfies the Kronecker delta property.

For cases other than the ones previously discussed such as a prescribed displacement imposed at interior control point 3 or a non-uniform Dirichlet BC[4] applied on edge BC, special treatment of Dirichlet BCs have to be employed as is the case for meshless methods. Techniques available include the Lagrange multiplier method, the penalty method, the augmented Lagrangian method and we refer to Nguyen et al. (2008) for an overview of these techniques in the context of meshless methods. In Wang and Xuan (2010) a transformation method was proposed to impose inhomogeneous

[4] Dirichlet BCs that vary from point to point.

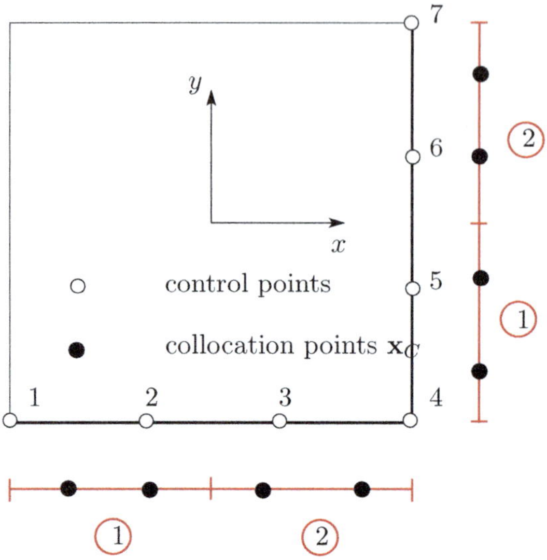

Figure 9: Illustration of the implementation of the least squares method: a quadratic NURBS surface with knot vectors $\Xi = \{0, 0, 0, 0.5, 1, 1, 1\}$ and $\mathscr{H} = \{0, 0, 0, 0.5, 1, 1, 1\}$. Essential BCs are imposed on the bottom and right edges. A one dimensional mesh for this boundary is created. The control points defining the essential boundary are numbered from one to the total number of boundary control points. Note that this numbering is required only for assembling the matrix $\mathbf{A}$.

Dirichlet BCs in IGAFEM. However this method requires modifications to the stiffness matrix which breaks the usual structure of a FE code. The authors in De Luycker et al. (2011) presented a weak enforcement of general inhomogeneous Dirichlet BCs using a least squares minimization and the implementation is described in what follows. The same procedure was used in imposing BCs in meshless methods when coupling a fluid to a solid domain through a master-slave concept Rabczuk et al. (2010). Imposing Dirichlet boundary conditions with Nitsche's method was presented in Embar et al. (2010) for spline-based finite elements.

The basic idea of the least squares method is to find the parameters of the boundary control points that minimize the following quantity

$$\begin{aligned} J &= \frac{1}{2}\sum_C ||\mathbf{u}(\mathbf{x}_C) - \bar{\mathbf{u}}(\mathbf{x}_C)||^2 \\ &= \frac{1}{2}\sum_C \left|\left|\sum_A \Phi_A(\mathbf{x}_C)\mathbf{q}_A - \bar{\mathbf{u}}(\mathbf{x}_C)\right|\right|^2, \end{aligned} \tag{43}$$

where $\mathbf{x}_C$ denotes a set of collocation points distributed on the essential boundary Γ_u, $\mathbf{q}_A$ are the parameters of the control points defining Γ_u and Φ_A represents the NURBS basis functions that are non-zero at $\mathbf{x}_C$. For the sake of clarity, let us consider the case where there is only one collocation point and a quadratic basis (thus there are 3 non-zero Φ_A at $\mathbf{x}_C$). So, we have

$$J = \frac{1}{2}\,||\Phi_1(\mathbf{x}_C)\mathbf{q}_1 + \Phi_2(\mathbf{x}_C)\mathbf{q}_2 + \Phi_3(\mathbf{x}_C)\mathbf{q}_3 - \bar{\mathbf{u}}(\mathbf{x}_C)||^2. \tag{44}$$

The partial derivatives of J with respect to $\mathbf{q}_i$ are given by

$$\begin{aligned} \frac{\partial J}{\partial \mathbf{q}_1} &= [\Phi_1(\mathbf{x}_C)\mathbf{q}_1 + \Phi_2(\mathbf{x}_C)\mathbf{q}_2 + \Phi_3(\mathbf{x}_C)\mathbf{q}_3 - \bar{\mathbf{u}}(\mathbf{x}_C)]\Phi_1(\mathbf{x}_C) \\ \frac{\partial J}{\partial \mathbf{q}_2} &= [\Phi_1(\mathbf{x}_C)\mathbf{q}_1 + \Phi_2(\mathbf{x}_C)\mathbf{q}_2 + \Phi_3(\mathbf{x}_C)\mathbf{q}_3 - \bar{\mathbf{u}}(\mathbf{x}_C)]\Phi_2(\mathbf{x}_C) \\ \frac{\partial J}{\partial \mathbf{q}_3} &= [\Phi_1(\mathbf{x}_C)\mathbf{q}_1 + \Phi_2(\mathbf{x}_C)\mathbf{q}_2 + \Phi_3(\mathbf{x}_C)\mathbf{q}_3 - \bar{\mathbf{u}}(\mathbf{x}_C)]\Phi_3(\mathbf{x}_C). \end{aligned} \tag{45}$$

The condition $\frac{\partial J}{\partial \mathbf{q}} = 0$ thus gives the following linear system

$$\begin{bmatrix} \Phi_1\Phi_1 & \Phi_2\Phi_1 & \Phi_3\Phi_1 \\ \Phi_1\Phi_2 & \Phi_2\Phi_2 & \Phi_3\Phi_2 \\ \Phi_1\Phi_3 & \Phi_2\Phi_3 & \Phi_3\Phi_3 \end{bmatrix}_{\mathbf{x}_C} \begin{bmatrix} q_1^x & q_1^y \\ q_2^x & q_2^y \\ q_3^x & q_3^y \end{bmatrix} = \begin{bmatrix} \bar{u}_x(\mathbf{x}_C)\Phi_1(\mathbf{x}_C) & \bar{u}_y(\mathbf{x}_C)\Phi_1(\mathbf{x}_C) \\ \bar{u}_x(\mathbf{x}_C)\Phi_2(\mathbf{x}_C) & \bar{u}_y(\mathbf{x}_C)\Phi_2(\mathbf{x}_C) \\ \bar{u}_x(\mathbf{x}_C)\Phi_3(\mathbf{x}_C) & \bar{u}_y(\mathbf{x}_C)\Phi_3(\mathbf{x}_C) \end{bmatrix}. \tag{46}$$

By collecting all the NURBS basis at point $\mathbf{x}_C$ in a column vector $\mathbf{N}(\mathbf{x}_C)$, the control points displacements in the x and y directions in $\mathbf{q}_x$ and $\mathbf{q}_y$, respectively, Equation (46) can be written in a compact form as

$$\begin{aligned} (\mathbf{N}(\mathbf{x}_C)\mathbf{N}(\mathbf{x}_C)^{\mathrm{T}})\mathbf{q}_x &= \bar{u}_x(\mathbf{x}_C)\mathbf{N}(\mathbf{x}_C) \\ (\mathbf{N}(\mathbf{x}_C)\mathbf{N}(\mathbf{x}_C)^{\mathrm{T}})\mathbf{q}_y &= \bar{u}_y(\mathbf{x}_C)\mathbf{N}(\mathbf{x}_C). \end{aligned} \tag{47}$$

Repeating the same analysis for other collocation points $\mathbf{x}_C$ on the Dirichlet boundary, one obtains the linear system $\mathbf{Aq} = \mathbf{b}$ with two different $\mathbf{b}$ (one for the x component and the other for the y component). The dimension of $\mathbf{A}$ is $n \times n$ where n denotes the number of control points defining the Dirichlet boundary. Having these boundary control point displacements, the enforcement of Dirichlet BCs (when solving $\mathbf{Ku} = \mathbf{f}$) are then treated as in standard FEM.

We note that this procedure involves only control points that define the essential boundary. This is in sharp contrast to meshless shape functions such as the MLS used in the Element Free Galerkin method Belytschko et al. (1994b) where the displacements at a point on the essential boundary depend not only on the nodes on that boundary but also the neighbouring interior nodes. It is said that NURBS therefore satisfy the so-called weak Kronecker delta property as local Maxent interpolants approximations.

Listing 3.1 gives the implementation of the least-squares method. We refer to Figure 9 for an illustration of this method.

Listing 3.1: Matlab® implementation of the least-squares method

```
A    = zeros(noDispNodes,noDispNodes);
bx   = zeros(noDispNodes,1);
by   = zeros(noDispNodes,1);
noxC    = 4; % number of collocation pts/element
% loop over bottom edge
for ie=1:noElemsU
    sctr    = bottomEdgeMeshIGA(ie,:);%connectivity array
    pts     = controlPts(sctr,:); % control pts of element ie
    sctrA   = bndElement(ie,:); %connectivity array for A matrix
    xiE     = elRangeU(ie,:); %parametric coords of ie
    xiArr   = linspace(xiE(1),xiE(2),noxC);%collocation pts x_C
    for ic=1:noxC
        xi                  = xiArr(ic);
        [N dNdxi]           = NURBS1DBasisDers(xi,p,uKnot,weights);
        A(sctrA,sctrA)  = A(sctrA,sctrA) + N'*N;
        x                   = N *pts;
        % exact displacements
        [ux,uy]    = ...
        bx(sctrA) = bx(sctrA) + ux*N';
        by(sctrA) = by(sctrA) + uy*N';
    end
end
% loop over other edges if neccessary
% solve the system Aq_x=bx and Aq_y=by
[LL UU]  = lu(A);
qxTemp   = LL\bx;
qyTemp   = LL\by;
qx       = UU\qxTemp;
qy       = UU\qyTemp;
% later, before solving Ku=f, qx and qy will be used
% to enforce BCs as in conventional FEM.
```

In our code implementation are provided for the penalty method, the Lagrange multiplier method and the least squares method. The implementation of the two former methods are considered standard and the reader is referred to Nguyen et al. (2008) for details.

4 MIGFEM- A Matlab IGA (X)FEM code

In this section we describe shortly the IGA Matlab (X)FEM code. The code supports one, two and three dimensional linear elasticity problems. Extended IGA for crack and material interface modelling is also implemented. Geometrically nonlinear solid and structural mechanics models (beams, plates and shells) are available. The features of the code, which is available at http://sourceforge.net/projects/cmcodes/, include:

- Global h,p and k-refinement is provided for one, two and dimensional meshes.
- Extended IGA which is a Partition of Unity enrichment IGA for 2D traction-free cracks and material interfaces is also implemented. Level sets are used to detect enriched nodes. However, enrichment functions are defined in terms of standard geometry i.e. not in terms of level sets.
- An ad hoc implementation for 3D cracks is also provided. In this case, level sets are used both for enrichment detection and enrichment function evaluation.
- Visualization of displacements and stresses are done in Paraview by exporting the results to a VTU file. Mesh for visualization purpose is Q4 in 2D and B8 in 3D where B8 denotes tri-linear brick elements.
- Inhomogeneous Dirichlet boundary conditions are treated with the penalty method, the Lagrange multiplier method and the least square method.
- Compatible multi-patch isogeometric formulation for two dimensional problems.
- Incompatible multi-patch isogeometric formulation for two/three dimensional problems.
- Support T-splines via the Bézier extraction operators.
- Structural mechanics including beams, plates and thin shells.
- Implicit Newmark scheme and explicit central difference scheme for time discretization.

4.1 Usage

Compiling the C files Before using the M files, you should first compile some C codes implemented as MEX files. Open Matlab and in the command line, type the following command "compile". Actually compile.m contains commands to compile MEX files.

Program structure The code is organized into sub-folders as explained in what follows.

iga contains the main files for linear elasticity problems e.g., *igaPlateCircularHole.m* for the plate with a circular hole problem. These are the main script files to start with.

xiga contains the main files for LEFM problems e.g., *xigaEdgeCrack.m* for the edged crack plate in tension problem.

data contains data files that specify control points, knots, basis orders and weights. The data files are named as "*Data.m".

C_files contains the C implementation of some NURBS algorithms (evaluation of NURBS basis and its derivatives etc.). In this way, MIGFEM is much faster than using Matlab NURBS implementation.

fem contains FEM M files to solve the same problems solved with IGA. Used as reference solution to check IGA implementation. Usually the FEM code reads GMSH mesh files (files with extension *.msh). The *.msh files are created from the GMSH geometry files *.geo.

fem-util contains some FEM related functions (Chessa (2002)) which are used in an IGA context (Gauss quadrature for instance).

@patch2D contains classes and methods implementation for multi-patch isogeometric analysis.

nurbs-geopdes is where the NURBS toolbox locates. We use the order elevation provided by this toolbox to perform k-refinement.

bezier-extraction contains an 2D implementation of NURBS-based FEM using the concept of Bézier extraction.

structural-mechanics contains implementation of various structural mechanics models including Euler-Bernoulli beam, Kirchhoff plate, Kirchhoff-Love thin shell (with and without geometrical nonlinearity). Mass matrices are also provided so that modal analysis and transient dynamics analysis can be performed.

geometric-nonlinear contains implementation of the Total Lagrangian formulation for large displacement and rotation solid mechanics problems.

results contain output files include postscript files (*.EPS) and VTU files (*.vtu). VTU files can be processed by Paraview to plot displacements and stresses contours.

4.2 Data structure

The MIGFEM follows the Matlab FEM code described in Chessa (2002). The main data structures include (1) *element* (store the element connectivity), (2) *node* (store nodal coordinates), (3) K (stiffness matrix) and (4) f (external force vector). In contrary to FEM in which the element connectivity and nodal coordinates are inputs which have been created by a meshing program, in IGA, the input consists of CAD data including knots, control points, order of basis functions. Therefore, one has to construct a FE mesh from those data.

Mesh generation Given the knot vectors $uKnot$ and $vKnot$ together with the orders of the basis p and q, one can compute the number of control points along ξ and ζ directions, denoted by n and m. Then we define a two dimensional matrix of dimension $m \times n$ called $node\_pattern$ given by

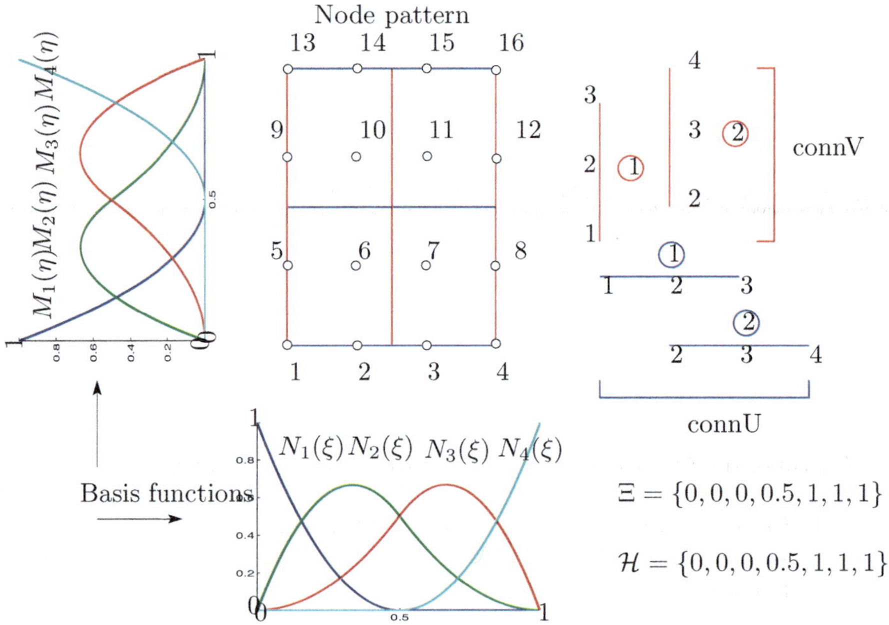

Figure 10: Two dimensional isogeometric analysis: mesh generation for a bi-quadratic NURBS surface (2×2 elements). The circles denote the control points.

$$node\_pattern = \begin{bmatrix} 1 & 2 & 3 & 4 \\ 5 & 6 & 7 & 8 \\ 9 & 10 & 11 & 12 \\ 13 & 14 & 15 & 16 \end{bmatrix} \tag{48}$$

for the case of there are 4 control points along ξ and 4 control points along η directions, see Figure 10. The number of elements along the two directions are $noElemsU$ = length(unique($uKnot$)) − 1 and $noElemsV$ = length(unique($uKnot$)) − 1. Then the connectivity matrix for the ξ direction, denoted by $connU$ which is a $noElemsU \times (p+1)$ matrix, is given by

$$connU = \begin{bmatrix} 1 & 2 & 3 \\ 2 & 3 & 4 \end{bmatrix}, \quad connV = \begin{bmatrix} 1 & 2 & 3 \\ 2 & 3 & 4 \end{bmatrix} \tag{49}$$

for the example under consideration. The matrix $connU$ stores the indices of the non-zero basis functions in every knot spans for the ξ direction. Having these information, we are able to define the connectivity matrix for the whole mesh, called $element$ which is a $noElemsU * noElemsV \times (p+1)(q+1)$ matrix. For the example herein, this matrix reads

$$element = \begin{bmatrix} 1 & 2 & 3 & 5 & 6 & 7 & 9 & 10 & 11 \\ 2 & 3 & 4 & 6 & 7 & 8 & 10 & 11 & 12 \\ 5 & 6 & 7 & 9 & 10 & 11 & 13 & 14 & 15 \\ 6 & 7 & 8 & 10 & 11 & 12 & 14 & 15 & 16 \end{bmatrix} \tag{50}$$

Finally in order to compute the mapping from the parent domain $\square$ to the parametric space, we define the range of every knot spans in two directions as

$$rangeU = \begin{bmatrix} 0 & 0.5 \\ 0.5 & 1 \end{bmatrix}, \quad rangeV = \begin{bmatrix} 0 & 0.5 \\ 0.5 & 1 \end{bmatrix} \tag{51}$$

In order to retrieve the parametric coordinates of a specific element, the matrix $index$, that is a $noElemsU * noElemsV \times 2$ matrix, is used for this purpose. For a given element e, its parametric coordinates are determined by *rangeU(index(e,1),:); rangeV(index(e,2),:)*.

The above discussion together with the illustration given in Figure 10 is implemented in Matlab in the file **generateIGA2DMesh.m** and shown in Listing 4.1.

Listing 4.1: Mesh generation for two dimensional problems.

```
uniqueUKnots   = unique(uKnot);
uniqueVKnots   = unique(vKnot);
noElemsU       = length(uniqueUKnots)-1;%#of elements xi dir.
noElemsV       = length(uniqueVKnots)-1;%#of elements eta dir.
nodePattern    = zeros(noPtsY,noPtsX);
count = 1;
for i=1:noPtsY
    for j=1:noPtsX
```

```
        nodePattern(i,j) = count;
        count            = count + 1;
    end
end
% determine our element ranges and the corresponding
% knot indices along each direction
[elRangeU,elConnU] = buildConnectivity(p,uKnot,noElemsU);
[elRangeV,elConnV] = buildConnectivity(q,vKnot,noElemsV);
noElems = noElemsU * noElemsV;
element = zeros(noElems,(p+1)*(q+1));
e = 1;
for v=1:noElemsV
    vConn = elConnV(v,:);
    for u=1:noElemsU
        c = 1;
        uConn = elConnU(u,:);
        for i=1:length(vConn)
            for j=1:length(uConn)
              element(e,c) = nodePattern(vConn(i),uConn(j));
              c = c + 1;
            end
        end
        e = e + 1;
    end
end
index = zeros(noElems,2);
count = 1;
for j=1:size(elRangeV,1)
    for i=1:size(elRangeU,1)
        index(count,1) = i;
        index(count,2) = j;
        count = count + 1;
    end
end
```

4.3 Assembly process

The assembly of an IGA-FEM code is given in Listing 4.2 where it can be seen that the procedure is almost identical to the conventional FEM. The minor differences lie in (1) the need of the elements in the parameter space (lines 4 to 7) and (2) the second map (from parent domain to parametric domain) in the numerical integration of the stiffness matrix (line 35).

Listing 4.2: Matlab code for IGA for 2D elasticity problems.

```
[W,Q]=quadrature(4, 'GAUSS', 2); % 4x4 point quadrature
% Assembling system of equation
for e=1:noElems % Loop over elements (knot spans)
  idu    = index(e,1);
  idv    = index(e,2);
  xiE    = elRangeU(idu,:); % [xi_i,xi_i+1]
  etaE   = elRangeV(idv,:); % [eta_j,eta_j+1]
  sctr   = element(e,:);    % element scatter vector
  sctrB  = [sctr sctr+noCtrPts]; % B scatter vec.
  nn     = length(sctr);
  B      = zeros(3,2*nn);
  % loop over Gauss points
  for gp=1:size(W,1)
    pt      = Q(gp,:);
    wt      = W(gp);
    % compute coords in parameter space
    Xi      = parent2ParametricSpace(xiE,pt(1));
    Eta     = parent2ParametricSpace(etaE,pt(2));
    J2      = jacobianPaPaMapping(xiE,etaE);
    % derivate of R=Nxi*Neta w.r.t xi and eta
    [dRdxi dRdeta] = NURBS2Dders([Xi; Eta],p,q,...
                                 uKnot,vKnot,weights');
    % Jacobian matrix
    pts         = controlPts(sctr,:);
    jacob       = pts'*[dRdxi' dRdeta'];
    J1          = det(jacob);
    invJacob    = inv(jacob);
    dRdx        = [dRdxi' dRdeta'] * invJacob;
    % B matrix
    B(1,1:nn)        = dRdx(:,1)';
    B(2,nn+1:2*nn)   = dRdx(:,2)';
    B(3,1:nn)        = dRdx(:,2)'; B(3,nn+1:2*nn)= dRdx(:,1)';
    % compute elementary stiffness matrix and
    % assemble it to the global matrix
    K(sctrB,sctrB) = K(sctrB,sctrB) + B'*C*B*J1*J2*wt;
  end
end
```

4.4 Post-processing

We present here a simple technique to visualize the IGA results that reuse available visualization techniques for finite elements. For the sake of simplicity, only 2D cases are considered here. In the first step, a mesh consisting of four-noded quadrilateral (Q4) elements is generated, see Figure 11. We call this mesh the visualization mesh (of which connectivity matrix is stored in *elementV*) whose nodes are images of the knots in the

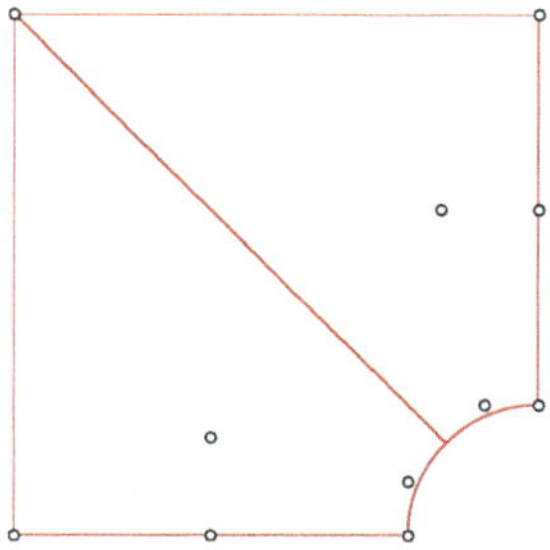

(a) NURBS mesh

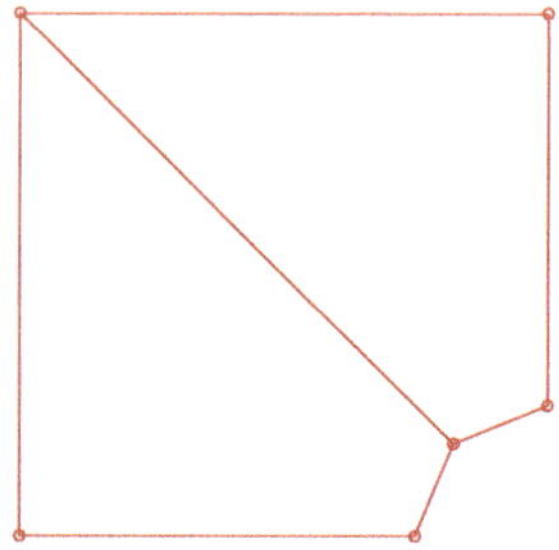

(b) approximate Q4 mesh

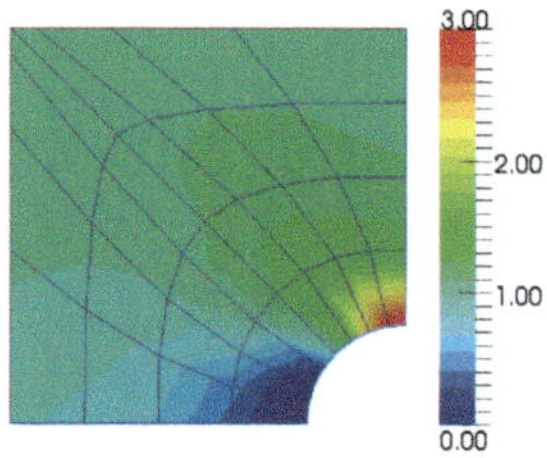

(c) stress visualization on a refined mesh

Figure 11: Exact NURBS mesh (top left) and approximate Q4 mesh (top right) for visualization purpose. The nodes in the Q4 mesh are the intersections of ξ and η knot lines. The bottom figure shows a contour plot of a stress field in Paraview. It should be emphasized that the mesh in (b) can not give a smooth contour plot. The result given in (c) was obtained with a refined NURBS mesh (hence a refined Q4 mesh).

physical space. In the second step, quantities of interest e.g., stresses are computed at the nodes of the Q4 mesh. This mesh together with the nodal values can then be exported to a visualization program such as Paraview, see Henderson (2007), for visualization. It should be emphasized that due to high order continuity of the NURBS basis, there is no need to perform nodal averaging as required in standard C^0 finite element analysis. Listing 4.3 gives the Matlab code for building the Q4 visualization mesh, computing the stresses at nodes of this mesh and Listing 4.3 is used to either visualize the stresses directly in Matlab or export the result to Paraview. The source code can be found in the file **plotStress1.m**. For NURBS solids, a mesh of brick elements is built. The script **plotStress3d.m** can be used for 3D visualization.

Listing 4.3: Matlab code for computation of stresses at nodes of the visualization mesh.

```
buildVisualizationMesh; % build visualization Q4 mesh
stress = zeros(noElems,4,3);
disp   = zeros(noElems,4,2);
for e=1:noElems
  idu    = index(e,1);
  idv    = index(e,2);
  xiE    = elRangeU(idu,:); % [xi_i,xi_i+1]
  etaE   = elRangeV(idv,:); % [eta_j,eta_j+1]
  sctr   = element(e,:);          % element scatter vector
  sctrB  = [sctr sctr+noCtrPts];%vector scatters B matrix
  nn     = length(sctr);
  B      = zeros(3,2*nn);
  pts    = controlPts(sctr,:);

  uspan = FindSpan(noPtsX-1,p,xiE(1),uKnot);
  vspan = FindSpan(noPtsY-1,q,etaE(1),vKnot);
  % loop over 4 nodes
  gp = 1;
  for iv=1:2
    if (iv==2) xiE = sort(xiE,'descend'); end
    for iu=1:2
      Xi  = xiE(iu);
      Eta = etaE(iv);
      [N dRdxi dRdeta] = NURBS2DBasisDersSpecial([Xi;Eta],...
          p,q,uKnot,vKnot,weights',[uspan;vspan]);
      % derivative w.r.t spatial physical coordinates
      jacob      = pts' * [dRdxi' dRdeta'];
      invJacob   = inv(jacob);
      dRdx       = [dRdxi' dRdeta'] * invJacob;
      % B matrix
      B(1,1:nn) = dRdx(:,1)'; B(2,nn+1:2*nn) = dRdx(:,2)';
      B(3,1:nn) = dRdx(:,2)'; B(3,nn+1:2*nn) = dRdx(:,1)';
      strain          = B*U(sctrB);
      stress(e,gp,:)  = C*strain;
      gp = gp +1;
    end
  end
end
```

Listing 4.4: Matlab code for post-processing.

```
% plot sigma_xx contour directly in Matlab
figure
plot_field(node,elementV,'Q4',stress(:,:,1));
% export to VTK format to plot in Mayavi or Paraview
sigmaXX = zeros(size(node,1),1);
sigmaYY = zeros(size(node,1),1);
sigmaXY = zeros(size(node,1),1);

dispX = zeros(size(node,1),1);
dispY = zeros(size(node,1),1);

for e=1:size(elementV,1)
    connect = elementV(e,:);
    for in=1:4
        nid = connect(in);
        sigmaXX(nid) = stress(e,in,1);
        sigmaYY(nid) = stress(e,in,2);
        sigmaXY(nid) = stress(e,in,3);
    end
end
VTKPostProcess(node,elementV,2,'Quad4','result.vtu',...
            [sigmaXX sigmaYY sigmaXY],[dispX dispY]);
```

For three-dimensional problems, the same procedure is used where a mesh of tri-linear brick elements is created and the values of interest are computed at the nodes of this mesh (see file **plotStress3d.m**). The results are then exported to Paraview under a structured grid format (*.vts files), see the file **mshToVTK.m**).

4.5 h, p, k-refinement

For the refinement of NURBS, we reuse the NURBS Toolbox described in de Falco et al. (2011). We construct a NURBS surface as shown in Figure 12a. The corresponding Matlab code is given in Listing 4.5. Using an uniform h-refinement (Listing 4.6), we obtain the mesh given in Figure 12b. If one needs to use $k-$refinement ($p-$refinement followed by $h-$refinement), then the code in Listing 4.7 can be used (see Figure 12c). Finally, Figure 12d gives the mesh which is obtained by the process in which $h-$refinement is employed first and then $p-$refinement is performed (see Listing 4.8). After using the NURBS toolbox, the NURBS object is then converted to MIGFEM data structures using the function **convert2DNurbs**.

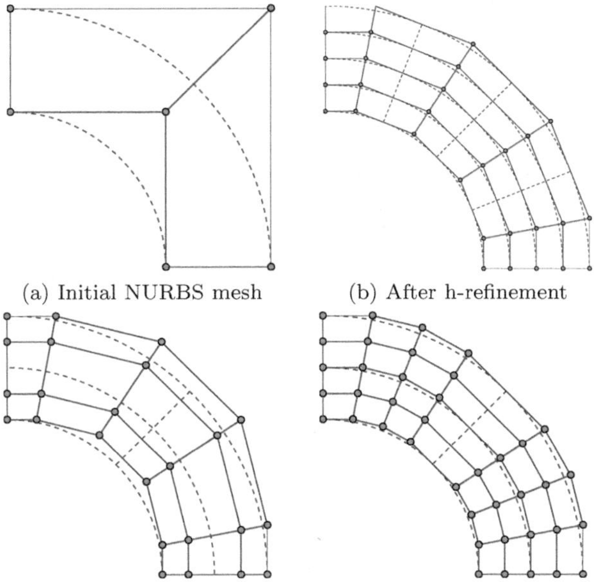

(a) Initial NURBS mesh (b) After h-refinement

(c) After k-refinement (d) After hp-refinement

Figure 12: Illustration of the utilization of the NURBS toolbox in building NURBS object (a). From the initial mesh, different meshes can be obtained using either $h-$refinement, $p-$refinement or combination thereof. As can be seen from (c) and (d), $k-$refinement (c) is more efficient than $hp-$refinement (d). The function **plotMesh.m** is used to plot NURBS mesh and control polygon.

Listing 4.5: Construct a NURBS surface using the NURBS toolbox.

```
a = 0.3; b = 0.5; % inner/outer radius
uKnot = [0 0 0 1 1 1]; %quadratic basis
vKnot = [0 0 1 1];      %linear basis
% homogeneous coords(x*w,y*w,z*w,w),3 pts u-dir , 2 pts v-dir
controlPts            = zeros(4,3,2);
controlPts(1:2,1,1)   = [a;0];
controlPts(1:2,2,1)   = [a;a;];
controlPts(1:2,3,1)   = [0;a];
controlPts(1:2,1,2)   = [b;0];
controlPts(1:2,2,2)   = [b;b];
controlPts(1:2,3,2)   = [0;b];
controlPts(4,:,:)     = 1;
controlPts(4,2,1)  = 1/sqrt(2);
controlPts(4,2,2)  = 1/sqrt(2);
% homogenous coordinates (x*w,y*w,z*w)
controlPts(1:2,2,1) = controlPts(1:2,2,1)*fac;
controlPts(1:2,2,2) = controlPts(1:2,2,2)*fac;
solid = nrbmak(controlPts,{uKnot vKnot});%build NURBS object
```

Listing 4.6: h-refinement using the NURBS toolbox.

```
refineLevel = 2;
for i=1:refineLevel
  uKnotVectorU = unique(uKnot);
  uKnotVectorV = unique(vKnot);
  % new knots along two directions
  newKnotsX =uKnotVectorU(1:end-1)+0.5*diff(uKnotVectorU);
  newKnotsY =uKnotVectorV(1:end-1)+0.5*diff(uKnotVectorV);

  newKnots  = {newKnotsX newKnotsY};
  solid     = nrbkntins(solid,newKnots);
  uKnot     = cell2mat(solid.knots(1));
  vKnot     = cell2mat(solid.knots(2));
end
convert2DNurbs % convert to the MIGFEM format
fi = 'mesh.eps';
plotMesh(controlPts,weights,uKnot,vKnot,p,q,res,'r-',fi);
```

Listing 4.7: k-refinement using the NURBS toolbox.

```
% code from Listing 5
% order elevation, p=p+2, q=q+1
solid = nrbdegelev(solid,[2 1]);
% then, knot insertion using the code from Listing 6
convert2DNurbs % convert to the MIGFEM format
```

Listing 4.8: h-refinement followed by p-refinement using the NURBS toolbox.

```
% code from Listing 5
% then, knot insertion using the code from Listing 6
% order elevation, p=p+2, q=q+1
solid = nrbdegelev(solid,[2 1]);
convert2DNurbs % convert to the MIGFEM format
```

Listing 4.9: A typical input file for MIGFEM.

```
% code from Listing 5
% code from Listing 6 (if only h-refinement)
convert2DNurbs%convert to the MIGFEM format (if not done yet)
generateIGA2DMesh % build the mesh (element connectivity)
```

4.6 Input file for MIGFEM

A typical input file is given in Listing 4.9. This file replaces the standard FE mesh file. Note that, for backward compatibility with older version of the code, some input files do not use the NURBS toolbox to create the NURBS object. For those input files, it is however impossible to perform order elevation and hence $k-$refinement. It is, therefore, recommended to create the NURBS objects using the NURBS toolbox which, besides the refinement functionalities, also supports many operations such as extrusion, rotation etc.

4.7 Bézier extraction

It has been recognized that B-splines/NURBS/T-splines basis can be written as a linear combination of the Bernstein polynomials. Mathematically,

$$\mathbf{N}^e(\boldsymbol{\xi}) = \mathbf{C}^e\mathbf{B}(\boldsymbol{\xi}) \tag{52}$$

where $\mathbf{C}^e$ denotes the elemental Bézier extraction operator and $\mathbf{B}(\boldsymbol{\xi})$ are the Bernstein polynomials which are defined on the parent element $[-1, 1] \times [1, 1]$. Since all elements have now the same set of the Bernstein basis, these basis can be pre-computed at integration points of a single parent element. Whenever a B-splines/T-splines shape function is needed, Equation (52) is then used that concerns only a matrix-vector multiplication in which $\mathbf{C}^e$ is sparse. Details and algorithms of Bézier extraction can be found in Borden et al. (2011); Scott et al. (2011).

Implementing IGA using the Bézier extraction has the following advantages especially when T-splines are employed

- A FE code capable of handling extraction operators can easily incorporate both NURBS and T-splines;
- It allows an analyst to do T-splines FEA without understanding the details of T-splines technology;
- Using the Bézier extraction, the NURBS/T-splines meshes contains elements having the same set of basis (the so-called Bernstein polynomials) defined on the parent domain as standard Lagrange finite elements.
- Simplify the parallelization of IGA code;
- It is faster than the IGA implementation without using the extraction operator (in our code the Bernstein polynomials are pre-computed at integration points of a single element).

Using the knots, the elemental (local) Bézier extraction operators can be computed using the file **bezierExtraction2D.m** for 2D and **bezierExtraction3D.m** for 3D problems. Listing 4.10 gives the implementation of the former. This is the additional step that one has to do. After having this data, there is no need to use knots and standard algorithms (find span for example) of NURBS to compute the shape functions. Listing 4.11 gives a part of the assembly of a 2D IGA code that uses the Bézier extraction. In order to compute the external force resulting from a traction being applied on an edge of a B-splines/NURBS/T-splines curve, one also needs the Bézier extraction operators along each parametric direction.

Listing 4.10: Computation of 2D Bézier extraction operators.

```
function [C,Cxi,Cet] = bezierExtraction2D(uknot,vknot,p,q)
% Bezier extraction operators for xi and eta
% nb1: number of elements along xi direction
% nb2: number of elements along eta direction
[Cxi,nb1] = bezierExtraction(uknot,p);
[Cet,nb2] = bezierExtraction(vknot,q);
% For Bsplines/NURBS, the element Bezier extraction
% operator is square.
size1 = size(Cxi(:,:,1),1);
size2 = size(Cet(:,:,1),1);
% Bezier extraction operators for the whole mesh
% as the tensor product of Cxi and Cet
C = zeros(size1*size2,size1*size2,nb1*nb2);
for eta=1:nb2
  for xi=1:nb1
    e = (eta-1)*nb1 + xi;
    for row=1:size2
      ird = (row-1)*size1 + 1;
      jrd =  row*size1;
      for col=1:size2
        icd = (col-1)*size1 + 1;
        jcd =  col*size1;
        C(ird:jrd,icd:jcd,e) = Cet(row,col,eta)*Cxi(:,:,xi);
      end
    end
  end
end
```

Listing 4.11: Assembly of stiffness matrix-IGA with Bézier extraction operators.

```
% pre-compute Bernstein polynomials
shapes   = zeros(noGpEle,noBasis);
derivs   = zeros(noGpEle,noBasis,2);

for gp=1:size(W,1)
  [shapes(gp,:) derivs(gp,:,:)]=...
                getShapeGradBernstein2D(p,q,Q(gp,1),Q(gp,2));
end
for e=1:noElems            % Loop over elements
     sctr     = element(e,:);              % element scatter vector
     Ce       = C(:,:,e);                  % element extraction op.
     we       = diag(weights(sctr,:));% element weights
     pts      = controlPts(sctr,:);    % element nodes
     Wb       = Ce'*weights(sctr,:);   % element Bezier weights
     for gp=1:size(W,1) % loop over Gauss points
         wt          = W(gp);
         % retrieve Bernstein basis and derivatives at GP gp
         Be          = shapes(gp,:)';
         dBedxi      = reshape(derivs(gp,:,:),noBasis,2);

         %% Bezier weight functions (denomenator of NURBS)
         wb             = dot(Be,Wb);
         wb2            = wb * wb;
         dwbdxi(1)  = dot(dBedxi(:,1),Wb);
         dwbdxi(2)  = dot(dBedxi(:,2),Wb);
         %% Shape function and derivatives
         R              =we*Ce*Be/wb;
         dRdxi(:,1)=we*Ce*(dBedxi(:,1)/wb-dwbdxi(1)*Be/(wb2));
         dRdxi(:,2)=we*Ce*(dBedxi(:,2)/wb-dwbdxi(2)*Be/(wb2));

         %% Jacobian matrix
         dxdxi  = pts'*dRdxi;dxidx=inv(dxdxi);dRdx=dRdxi*dxidx;
         detJ   = det(dxdxi);
         % and B matrix (use dRdx), K matrix as usual
```

4.8 More on Bernstein functions and Bézier extraction

In this section more details on Bernstein functions and the Bézier extraction are provided based on Borden et al. (2011); Scott et al. (2011). The univariate Bernstein basis functions of order p are defined over the biunit interval $[-1, 1]$ as

$$B_{i,p}(\xi) = \frac{1}{2^p}\binom{p}{i-1}(1-\xi)^{p-(i-1)}(1+\xi)^{i-1} \tag{53}$$

where the binomial coefficient $\binom{p}{i-1} = \frac{p!}{(i-1)!(p+1-i)!}$, $1 \leq i \leq p+1$. We emphasize that, in CAD Bernstein polynomials are defined in the interval $[0, 1]$. However, in a finite element setting, the biunit interval $[-1, 1]$, where the Gauss quadrature is defined, is preferable. The Bernstein basis functions for $p = 1, 2, 3$ and 4 are plotted in Figure 13. Note that for $p = 1$, the Bernstein basis resemble the linear Lagrange basis.

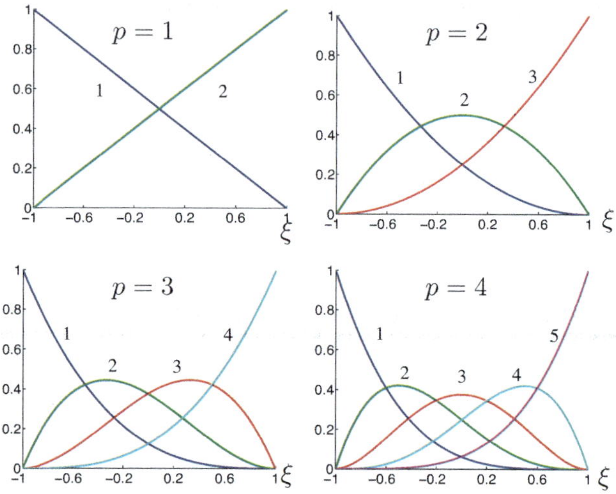

Figure 13: Bernstein basis functions for polynomial degree $p = 1, 2, 3, 4$.

The first derivative of the Bernstein basis is defined in terms of low order basis as follows

$$\frac{\partial B_{i,p}(\xi)}{\partial \xi} = \frac{1}{2} p \left[B_{i-1,p-1}(\xi) - B_{i,p-1}(\xi) \right] \tag{54}$$

And the second derivative is therefore given by

$$\frac{B^2_{i,p}(\xi)}{\partial \xi^2} = \frac{1}{4} p(p-1) \left[B_{i-2,p-2} - 2B_{i-1,p-2} + B_{i,p-2} \right] \tag{55}$$

The univariate Bernstein basis and its first derivatives are implemented in **getShapeGradBernstein.m**. The univariate Bernstein basis, its first and second derivatives are implemented in **getShapeGrad2Bernstein.m**. The bivariate Bernstein basis and its first derivatives are implemented in **getShapeGradBernstein2D.m** and **getShapeGrad2Bernstein2D.m** computes the basis, first and second derivatives. Trivariate Bernstein polynomials are implemented in **getShapeGrads3D.m**. Note that for 3D cases, only first derivatives are currently provided.

Given a B-spline curve of order p, and a knot vector, additional knots may be inserted at the internal knots by the use of knot insertion, until the multiplicity of each knot equals p. By doing so, the B-spline basis functions will be C^0-continuous between each element, and within each element they will be identical to the Bernstein polynomials of order p. This series of knot insertions is called Bézier decomposition. Figure 14 gives an example of this process. The figures were created using the file **bezier-extraction-example.m**.

The Bézier control points and the original control points are related to each other via the following equation

$$\mathbf{P}^b = \mathbf{C}^{\mathrm{T}}\mathbf{P} \tag{56}$$

The B-spline curve written in a matrix form is given by

$$\mathbf{x}(\xi) = \mathbf{P}^{\mathrm{T}}\mathbf{N}(\xi) \tag{57}$$

And this curve in terms of Bézier points is written as

$$\mathbf{x}(\xi) = (\mathbf{P}^b)^{\mathrm{T}}\mathbf{B}(\xi) = \mathbf{P}^{\mathrm{T}}\mathbf{C}\mathbf{B}(\xi) \tag{58}$$

where Equation (56) was used. By comparing the two Equations (57) and (58), one obtains

$$\mathbf{N}(\xi) = \mathbf{C}\mathbf{B}(\xi) \tag{59}$$

which means that the B-splines basis can be expressed as a linear combination of the Bernstein polynomials. In what follows we are going to extend this result to NURBS.

A NURBS curve can be written as follows by using a matrix notation

$$x(\xi) = \frac{1}{W(\xi)}\mathbf{P}^{\mathrm{T}}\mathbf{W}\mathbf{N}(\xi) = \frac{1}{W(\xi)}\mathbf{P}^{\mathrm{T}}\mathbf{W}\mathbf{C}\mathbf{B}(\xi) \tag{60}$$

in which use was made of Equation (59) in the second equality. Next, we are going to rewrite the weight function $W(\xi)$ in terms of the Bernstein polynomials as follows

$$W(\xi) = \mathbf{w}^{\mathrm{T}}\mathbf{N}(\xi) = \mathbf{w}^{\mathrm{T}}\mathbf{C}\mathbf{B}(\xi) \equiv (\mathbf{w}^b)^{\mathrm{T}}\mathbf{B}(\xi) \tag{61}$$

in which $\mathbf{w}^b$ are the so-called Bézier weights which are given by

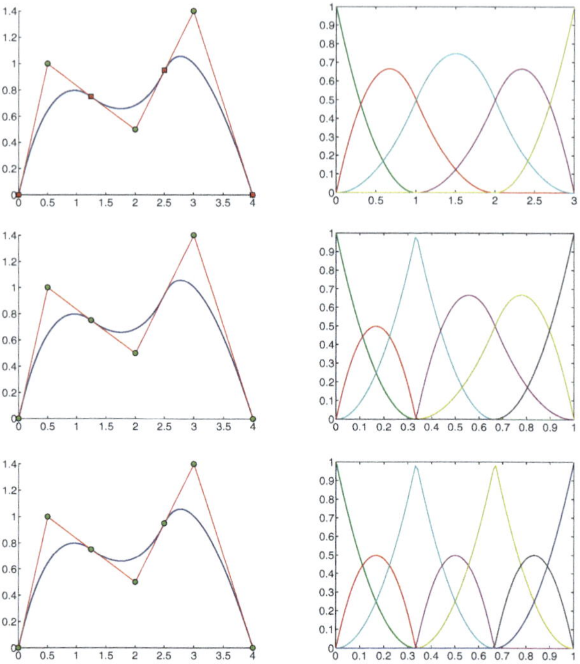

(a)

(b)

(c)

Figure 14: Illustration of the Bézier extraction concept on a quadratic B-splines curve with knot given by $H = [0, 0, 0, 1, 2, 3, 3, 3]$. The original curve and basis are given in first row (a). There are 5 control points/basis functions. The curve consists of three segments (in terms of FEM, there are three elements) and there is no control points at the interior element boundaries. Next, we insert 1 to the knot to get $H_1 = [0, 0, 0, 1, 1, 2, 3, 3, 3]$ and the new curve and basis functions are given in the second row (b)–there are now 6 points/basis. Finally, we add 2 to the knot to get $H_2 = [0, 0, 0, 1, 1, 2, 2, 3, 3, 3]$ and the new curve and basis are given in the third row (c). There are now 7 control points/basis. Each segment is now approximated by the same basis functions–the so-called Bernstein polynomials (cf. Figure 13 for $p = 2$).

$$\mathbf{w}^b = \mathbf{C}^{\mathrm{T}}\mathbf{w} \tag{62}$$

The same NURBS curve, which is now written in terms of Bézier control points/weights and Bernstein polynomials, is given by

$$x(\xi) = \frac{1}{W(\xi)}(\mathbf{P}^b)^{\mathrm{T}}\mathbf{W}^b\mathbf{B}(\xi) \tag{63}$$

where $\mathbf{W}^b$ is a diagonal matrix containing the Bézier weights. By comparing the two Equations (60) and (63), one obtains the expression to compute the Bézier NURBS control points from the NUBRS points

$$\mathbf{P}^b = (\mathbf{W}^b)^{-1}\mathbf{C}^{\mathrm{T}}\mathbf{W}\mathbf{P} \tag{64}$$

Note that matrices $\mathbf{W}$ and $\mathbf{W}^b$ are diagonal matrices hence invariant to the transpose operator.

The bivariate version of Equation (52) is given by

$$\mathbf{N}^e(\xi,\eta) = \mathbf{C}^e(\xi) \otimes \mathbf{C}_e(\eta)\mathbf{B}(\xi,\eta) \tag{65}$$

where $\mathbf{C}^e(\xi)$ and $\mathbf{C}^e(\eta)$ are the univariate elemental Bézier extractors in the ξ and η directions, respectively; $\mathbf{B}(\xi,\eta)$ denote the bivariate Bernstein basis functions. The tensor product operator is $\otimes$ defined as

$$\mathbf{A} \otimes \mathbf{B} = \begin{bmatrix} A_{11}\mathbf{B} & A_{12}\mathbf{B} & \cdots A_{1n}\mathbf{B} \\ A_{21}\mathbf{B} & A_{22}\mathbf{B} & \cdots A_{2n}\mathbf{B} \\ & & \\ A_{n1}\mathbf{B} & A_{n2}\mathbf{B} & \cdots A_{nn}\mathbf{B} \end{bmatrix} \tag{66}$$

where $\mathbf{A}$ and $\mathbf{B}$ can be of different sizes.

Extension to trivariate NURBS is straightforward due to the tensor product nature, one thus can write

$$\begin{aligned} \mathbf{N}^e(\xi,\eta,\zeta) &= \mathbf{C}^e(\xi) \otimes \mathbf{C}^e(\eta) \otimes \mathbf{C}^e(\zeta)\mathbf{B}(\xi,\eta,\zeta) \\ &= [\mathbf{C}^e(\xi,\eta) \otimes \mathbf{C}^e(\zeta)]\mathbf{B}(\xi,\eta,\zeta) \end{aligned} \tag{67}$$

in which the second equation is used for the implementation.

Evaluation of basis and its derivatives For ease of computer implementation we provide expressions for evaluating the shape functions, its first and second derivatives in the case of bi-variate NURBS implemented using the Bézier extraction. The shape functions for element e read

$$\mathbf{R}^e(\boldsymbol{\xi}) = \frac{\mathbf{W}^e\mathbf{N}^e(\boldsymbol{\xi})}{W(\boldsymbol{\xi})} = \frac{\mathbf{W}^e\mathbf{C}^e\mathbf{B}^e(\boldsymbol{\xi})}{W(\boldsymbol{\xi})} \tag{68}$$

where $\mathbf{W}^e$ denotes the diagonal matrix of weights and $W(\boldsymbol{\xi}) = \mathbf{w}^b \cdot \mathbf{B}^e$.

The derivatives of the shape functions with respect to ξ are given by

$$\mathbf{R}^e_{,\xi}(\boldsymbol{\xi}) = \mathbf{W}^e\mathbf{C}^e \left[\frac{\mathbf{B}^e_{,\xi}}{W} - \frac{\mathbf{B}^e W_{,\xi}}{W^2}\right] \tag{69}$$

And the second derivatives read

$$\mathbf{R}^e_{,\xi\xi} = \mathbf{W}^e\mathbf{C}^e \left[\frac{\mathbf{B}^e_{,\xi\xi}}{W} - \frac{2\mathbf{B}^e_{,\xi}W_{,\xi}}{W^2} - \frac{\mathbf{B}^e W_{,\xi\xi}}{W^2} + \frac{2\mathbf{B}^e(W_{,\xi})^2}{W^3}\right] \tag{70}$$

$$\mathbf{R}^e_{,\xi\eta} = \mathbf{W}^e\mathbf{C}^e \left[\frac{\mathbf{B}^e_{,\xi\eta}}{W} - \frac{\mathbf{B}^e_{,\xi}W_{,\eta}}{W^2} - \frac{\mathbf{B}^e_{,\eta}W_{,\xi}}{W^2} - \frac{\mathbf{B}^e W_{,\xi\eta}}{W^2} + \frac{2\mathbf{B}^e W_{,\xi}W_{,\eta}}{W^3}\right] \tag{71}$$

4.9 T-splines surfaces

Thanks to the implementation of IGA using the Bézier extraction operators, MIGFEM is able now to do FEA on T-splines geometries. Firstly, users have to build T-splines geometries using Rhino3d, then use the T-splines plugin to create the so-called '*.iga' files that are ready for being used in a IGA code. We reuse the script of GeoPDEs de Falco et al. (2011) that reads the '*.iga' files. This file contains all information that a IGAFEM code needs–element connectivity, nodal coordinates, weights and the Bézier extraction operators. We refer to Scott et al. (2011) for a description of Bézier extraction of T-splines. Comparing to the implementation of NURBS-based IGA (using Bézier extraction), as given in Listing 4.11, the implementation of T-splines-based IGA is almost the same except that data structure for storing the elemental Bézier extraction operators $\mathbf{C}$ and the element connectivity should be changed (in Matlab cell data structure should be used instead of matrix) since in T-splines, the number of non-zero basis function varies from elements to elements. We refer to Listing 4.12 for an implementation example. Furthermore, in unstructured T-splines meshes, there are elements of different orders and this should be taken into account. We refer to file **igaSquareTspline.m** for an implementation for unstructured Tspline meshes.

Listing 4.12: IGA with Bézier extraction for T-splines.

```
filename = 'plate_with_hole.iga';
tspline    = read_bezier_extraction (filename);
convert2DTsplines;
for e=1:noElems
   sctr    = element{e};             % element scatter vector
   sctrB   = [sctr sctr+noCtrPts];
   nn      = length(sctr);
   % Bezier extractors stored in a cell not a matrix!!!
   Ce      = C{e};                   % element Bezier ext. op.
   we      = diag(weights(sctr));    % element weights
   pts     = controlPts(sctr,:);     % element nodes
   Wb      = Ce'*weights(sctr);      % element Bezier weights
end
```

4.10 Post-processing

When one implements IGA using the Bézier extraction operators, the post-processing described in Section 4.4 can no longer be used. This is particularly true for T-splines. Furthermore, the previously presented post-processing method does not reuse the quantities at the Gauss points (e.g., stresses, strains). In what follows, some straightforward post-processing techniques that work for NURBS/T-splines are discussed.

For stresses and strains which are Gauss point (GP) quantities, one can use a discrete visualization technique listed in Listing 4.13. Listing 4.14 presents a method that requires a triangulation of the global Gauss points. In this way, the GPs serve as nodes and its triangulation as the elements. Therefore, available visualization for FEM (e.g., line 4 in Listing 4.14) can be readily reused. The drawback of this triangulation technique is that sometimes one has to remove undesired triangles which locate outside the physical domain. Note that the Gauss points constitute a mesh of unconnected quadrilateral elements (see Figure 15c). In case that there are many GPs inside each elements, the gap between elements will be reduced and a nice visualization can be obtained. This is the technique used in the GeoPDEs package.

Listing 4.13: Discrete visualization technique.

```
% x denotes the Gauss points in the physical domain
% sigma(:,2): sigma_{yy} at all GPs
figure
scatter(x(:,1),x(:,2),80,sigma(:,2),'full');
axis('equal');
colorbar
```

Listing 4.14: Finite element visualization using a triangulation of the Gauss points.

```
% x denotes the Gauss points in the physical domain
tri = delaunay(x);
figure
plot_field(x,tri,'T3',sigma(:,2));
axis('equal');
colorbar('vert');
```

Yet another method is to extrapolate the quantities at GPs (stresses and strains) to the Bézier control points, see Figure 16. Next, each Bézier element is divided into $n \times n$ sub-elements (n denotes the Bernstein degree). Lastly, standard FEM visualization methods can be employed on the mesh consisting of those sub-elements. The file **igaPlateCircularHoleTspline.m** implements all these techniques (except the last that uses sub-elements).

5 Extended isogeometric finite element method

Discrete cracks have been incorporated in a finite element (FE) context using zero-thickness interface elements, see for example Alfano and Crisfield (2001); Allix et al. (1995); Camacho and Ortiz (1996); Espinosa et al. (1998); Kerfriden et al. (2009); Nguyen (2014); Pandolfi and Ortiz (2002); Park et al. (2012); Schellekens and Borst (1993); Xu and Needleman (1994); Zhang et al. (2007b), elements with embedded discontinuities Armero and Linder (2009); da Costa et al. (2009); Simo et al. (1993) and elements with discontinuous enrichment via the extended/generalized finite element method (XFEM/GFEM) Moës et al. (1999); Strouboulis et al. (2001). A comparative study on the modelling of quasi-static discontinuous fracture using these techniques was given in Dias-da Costa et al. (2010); of dynamic fracture in Song et al. (2008) and a review of computational methods for fracture in quasi-brittle solids has recently been reported in Rabczuk (2013). By using

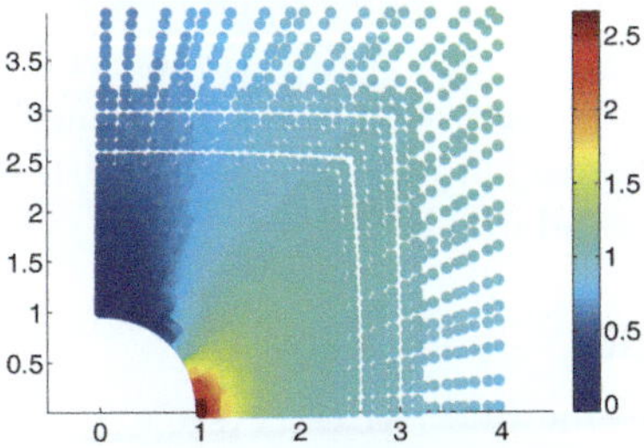

(a) Discrete visualization

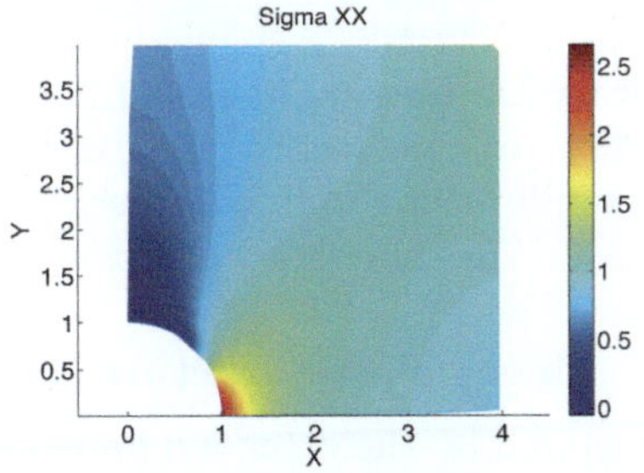

(b) Domain visualization with triangles

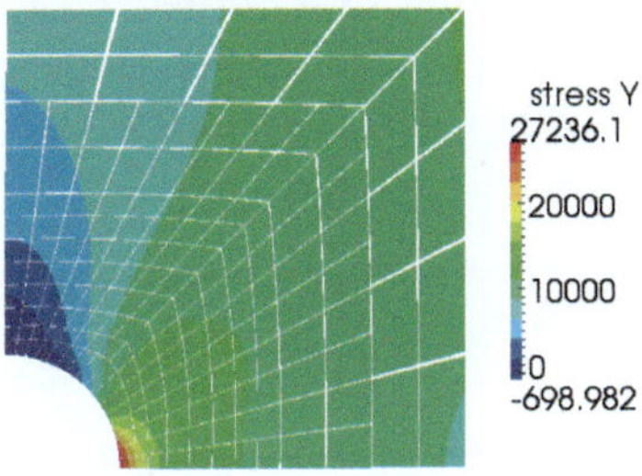

(c) Domain visualization with rectangles

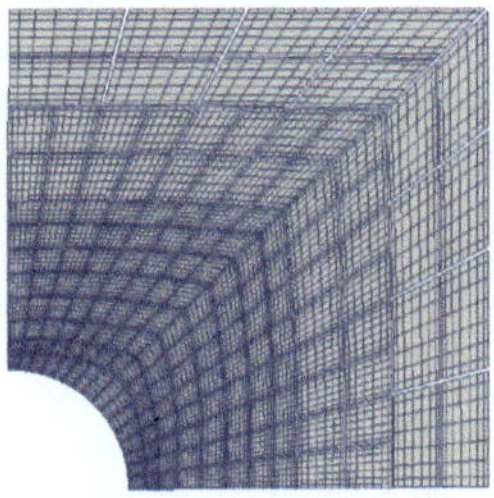

(d) Unconnected sub-elements

Figure 15: Simple visualization techniques that are available in MIGFEM: visualization of T-splines (c,d) using Paraview.

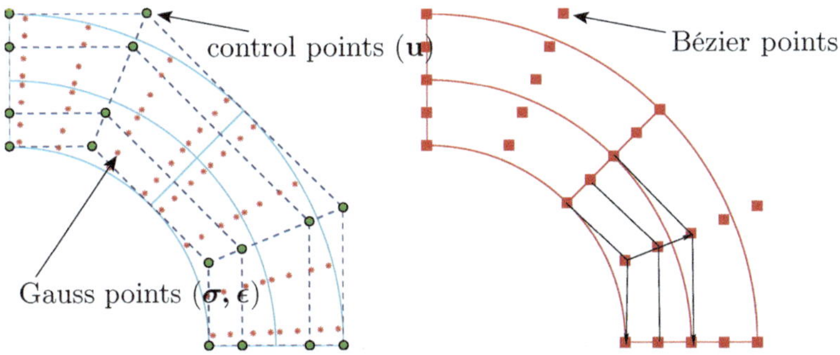

Figure 16: IGA visualization using the Bézier control mesh. The stresses (strains) at Gauss points are extrapolated to the Bézier control points. The displacements at NURBS/T-splines control points are mapped to the Bézier control points. Lastly, standard visualization techniques can be applied to the Bézier control mesh which is simply a mesh of quadrilateral elements in 2D.

B-splines/NURBS/T-splines in the XFEM one can model fracture with high accurary De Luycker et al. (2011); Ghorashi et al. (2012); Verhoosel et al. (2011a). In this section the extended isogeometric finite element method with applications to linear elastic fracture mechanics is presented with emphasis on the differences compared to a standard XFEM.

5.1 Linear elastic fracture mechanics

Inspired by the extended finite element method (XFEM) (see e.g., Moës et al. (1999) and Fries and Belytschko (2010) for a recent review), an extended isogeometric finite element formulation (XIGA) has been presented in De Luycker et al. (2011); Ghorashi et al. (2012) in which the displacement field is approximated as follows for a solid with traction-free cracks

$$\mathbf{u}^h(\mathbf{x}) = \sum_{I\in\mathcal{S}} \Phi_I(\mathbf{x})\mathbf{u}_I + \sum_{J\in\mathcal{S}^c} \Phi_J(\mathbf{x})H(\mathbf{x})\mathbf{a}_J + \sum_{K\in\mathcal{S}^f} \Phi_K(\mathbf{x})\left(\sum_{\alpha=1}^{4} B_\alpha \mathbf{b}_K^\alpha\right) \tag{72}$$

where Φ_I are NURBS basis functions. In addition to the standard dofs $\mathbf{u}_I$, there are additional dofs $\mathbf{a}_J$ and $\mathbf{b}_K^\alpha$. The Heaviside function is given by

$$H(\mathbf{x}) = \begin{cases} +1 & \text{if } (\mathbf{x}-\mathbf{x}^*)\cdot\mathbf{n} \geq 0 \\ -1 & \text{otherwise} \end{cases} \tag{73}$$

where $\mathbf{x}^*$ is the projection of point $\mathbf{x}$ on the crack. And the branch functions are given by

$$[B_1, B_2, B_3, B_4] = \left[\sqrt{r}\sin\frac{\theta}{2}, \sqrt{r}\cos\frac{\theta}{2}, \sqrt{r}\sin\frac{\theta}{2}\cos\theta, \sqrt{r}\cos\frac{\theta}{2}\cos\theta\right] \tag{74}$$

where r and θ are polar coordinates in the local crack front coordinate system. The set $\mathcal{S}^c$ includes the nodes whose support is cut by the crack whereas the set $\mathcal{S}^f$ are nodes whose support contains the crack tip $\mathbf{x}_{tip}$, see Figure 17.

Using the standard Bubnov-Galerkin procedure, the usual discrete equations, $\mathbf{Ku} = \mathbf{f}$ are obtained with only one difference in the $\mathbf{B}$[5] matrix which is now larger

$$\mathbf{B} = \left[\ \mathbf{B}^{\text{std}} \mid \mathbf{B}^{\text{enr}}\ \right] \tag{75}$$

where $\mathbf{B}^{\text{std}}$ is the standard $\mathbf{B}$ and $\mathbf{B}^{\text{enr}}$ is the enriched $\mathbf{B}$ matrix:

[5] with the assumption that nodes on essential are not enriched.

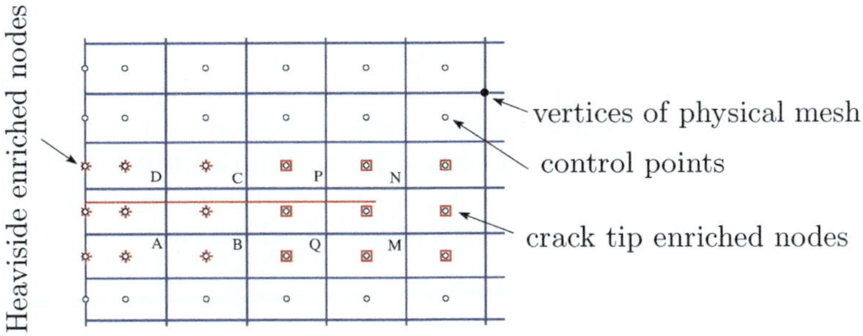

Figure 17: Illustration of enriched node sets $\mathcal{S}^f$ and $\mathcal{S}^c$ for a quadratic NURBS mesh. The read thick line denotes the crack.

$$\mathbf{B}_I^{\text{enr}} = \begin{bmatrix} (\Phi_I)_{,x}\Psi_I + \Phi_I(\Psi_I)_{,x} & 0 \\ 0 & (\Phi_I)_{,y}\Psi_I + \Phi_I(\Psi_I)_{,y} \\ (\Phi_I)_{,y}\Psi_I + \Phi_I(\Psi_I)_{,y} & (\Phi_I)_{,x}\Psi_I + \Phi_I(\Psi_I)_{,x} \end{bmatrix} \tag{76}$$

where $\Psi_I(\mathbf{x})$ can be either the Heaviside function $H(\mathbf{x})$, or the branch functions $B_\alpha(\mathbf{x})$. The unknowns vector $\mathbf{u}$ contains both displacements and enriched dofs.

The difference of an XFEM implementation compared to an FEM one are as follows

- Selection of enriched nodes;
- Numerical integration of elements intersecting the crack;
- Post-processing (J-integral computation) to extract the stress intensity factors;
- Crack visualization.

There are some minor modifications to these items when one implements XFEM into an IGAFEM code. In what follows, these issues are presented. This extended IGAFEM can be implemented within an available IGAFEM code with little modification following the ideas given in Nguyen et al. (2008). Briefly speaking, fictitious control points are added to handle additional dofs $\mathbf{a}$ and $\mathbf{b}$. A generic implementation of XFEM (in C++ programming language) has been presented in Bordas et al. (2007).

Selection of enriched nodes We herein present a way how to select enriched nodes that reuse all the related XFEM code. We use the approximate Q4 mesh used for visualization as discussed in Section 4.4 for selection of enriched control points. The level set values of the crack at the vertices of this mesh are fisrtly computed. Based on these level sets, elements cut by the crack and element containing the crack tip can be determined Sukumar et al. (2000). For example, element ABCD in Figure 17, is cut by the crack. In a FEM context, its four nodes are then Heaviside enriched. In an isogeometric framework, however, the control points associated to this element are enriched. Listing 5.1 is the Matlab implementation of this process. Note that this list is completely copied from our XFEM code with only minor modifications at lines 13 and 17 where sctrIGA was used instead of sctr.

Listing 5.1: Selection of enriched nodes.

```
enrich_node = zeros(noCtrPts,1);
count1 = 0;
count2 = 0;
for iel = 1 : numelem
    sctr    = elementV(iel,:);
    sctrIGA = element(iel,:);
    phi  = ls(sctr,1); % normal level set
    psi  = ls(sctr,2); % tangent level set
    if ( max(phi)*min(phi) < 0 )
        if max(psi) < 0
            count1 = count1 + 1 ; % one split element
            split_elem(count1) = iel;
            enrich_node(sctrIGA)   = 1;
        elseif max(psi)*min(psi) < 0
            count2 = count2 + 1 ; % one tip element
            tip_elem(count2) = iel;
            enrich_node(sctrIGA)   = 2;
        end
    end
end
split_nodes = find(enrich_node == 1);
tip_nodes   = find(enrich_node == 2);
```

Computation of stress intensity factors (SIFs) The mixed-mode SIFs are computed using the interaction integral and the implementation is standard. It should be emphasized, however, that in the computation of the interaction integral, we use bilinear shape functions (of Q4 finite element), in regardless of the adopted NURBS basis order for the geometry and displacement field, to compute the derivatives of the weight function. This guarantees that the weight function takes a value of unity on an open set containing the crack tip and vanishes on an outer contour as shown in Figure 18.

Numerical integration For elements intersecting with the crack, the sub-triangulation technique, as is standard in XFEM, is adopted without any

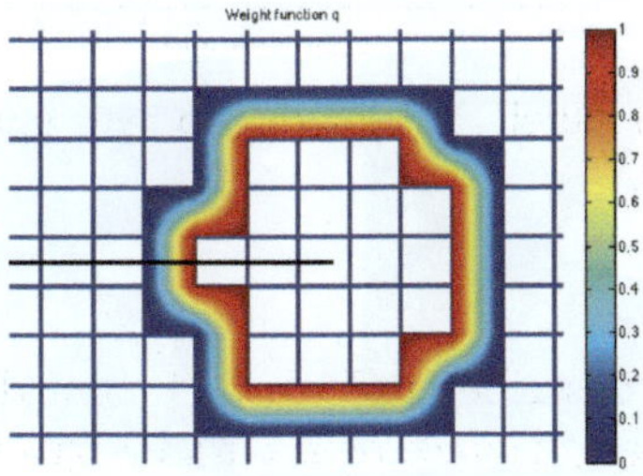

Figure 18: Distribution of weight function used in the computation of the interaction integral. Four-noded quadrilateral elements with bilinear Lagrange shape functions are used for the weight function.

changes.

Crack visualization is needed if a deformed mesh with a physical crack, instead of severely stretched elements, is required. We have $H(\mathbf{x}^+)-H(\mathbf{x}^-) = 2$ and $B_1(\mathbf{x}^+) - B_1(\mathbf{x}^-) = 2\sqrt{r}$, therefore the displacement jump at a point $\mathbf{x}$ on the crack face is given by

$$[\![\mathbf{u}]\!](\mathbf{x}) = 2 \sum_{J \in \mathcal{S}^c} \Phi_J(\mathbf{x})\mathbf{a}_J + 2\sqrt{r} \sum_{K \in \mathcal{S}^f} \Phi_K(\mathbf{x})\mathbf{b}_K^1 \tag{77}$$

Note that other branch functions are continuous functions and thus do not contribute to the displacement jump.

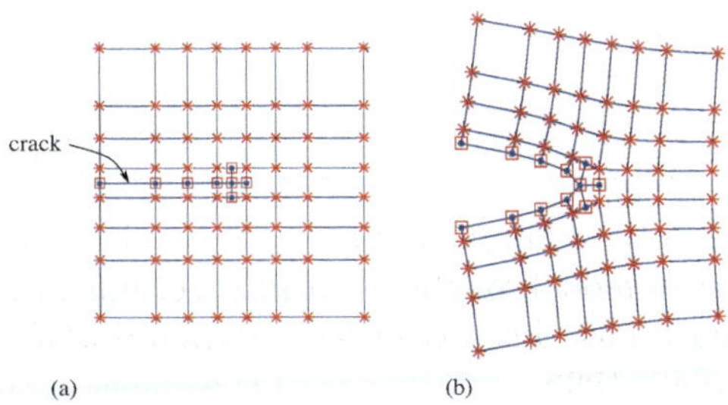

Figure 19: Crack visualization in XIGA: (a) building mesh that is compatible to the crack by putting double nodes along the crack (square nodes) and (b) the displacement jumps are assigned to these new nodes. Note that the square nodes in the tip element are only for compatibility purpose.

Figure 20: Contour plots on a cracked mesh: (a) vertical displacement and (b) normal stress in the vertical direction.

Figure 19 illustrates the idea of crack visualization and the corresponding implementation can be found in script **crackedMeshNURBS.m**. The contour plots of the displacement and stress field of a mode I cracked sample are given in Figure 20. Note that the stresses at points on the crack are simply set to zeros (traction-free cracks) and the stresses of the new nodes of the tip element are interpolated from the values of the four nodes of this Q4 element.

5.2 Voids

Although modelling voids does not require any enrichment, the idea of discretizing the internal geometries independent of the mesh shares similarity with the PUM based enriched methods. For this reason, we present here a method to model voids with NURBS based elements. The scheme is illustrated in Figure 21. There are three types of elements namely *(i)* standard elements which do not intersect with the voids and they are treated as usual *(ii)* voids elements which are simply neglected during the assembly process and *(iii)* split elements–elements cut by the voids which require a special integration scheme to be discussed in the sequel.

Integration on split elements Integration on elements cut by the voids can be achieved using either the sub-triangulation method or the adaptive Gauss rule, see Figure 22. In the latter, the Q4 element in the parent domain is firstly divided into four equal sub-cells. If a cell of these four cells is cut by the void, it is subdivided into four sub-cells. The process continues until a predefined maximum level is attained. This results in an aggregation of Gauss points around the boundary of the void. Figures 23 and 24 presents an example with a maximum level of three.

In case of circular voids, the check of whether a subcell is cut by the

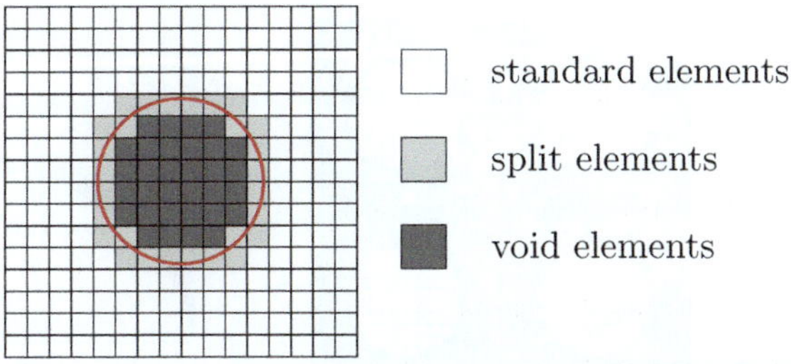

Figure 21: Void modelling on a structured mesh which is independent of the void geometry. Elements not cut by the voids (standard elements) are treated as conventional elements. Elements in the void space (void elements) are skipped in the assembly process and elements intersect with the voids (split elements) require special integration.

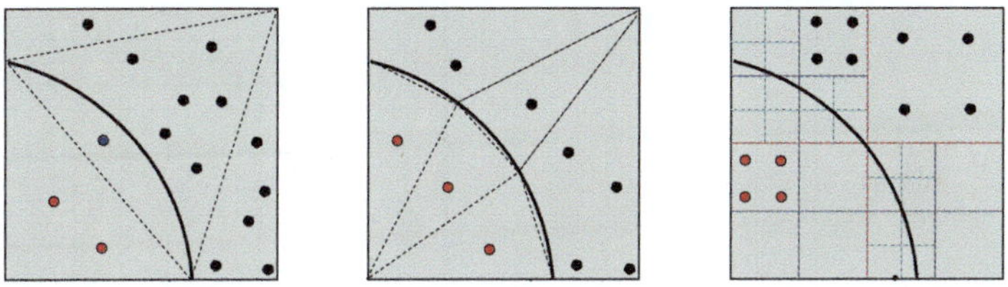

(a) sub-triangulation (ST) (b) refined ST (c) adaptive Gauss rule

Figure 22: Numerical integration on split elements: (a) sub-triangulation method (b) ST method with more sub-triangles to better capture the void boundary and (c) adaptive Gauss rule. Note that Gauss points locate in the void space (denoted by red circles) are deleted.

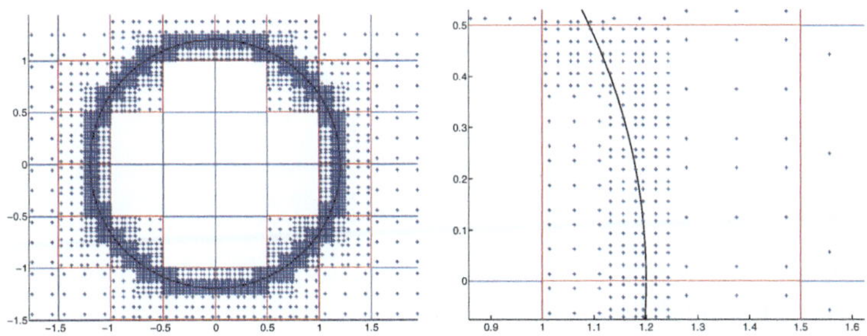

Figure 23: Numerical integration on split elements using the adaptive Gauss rule in action.

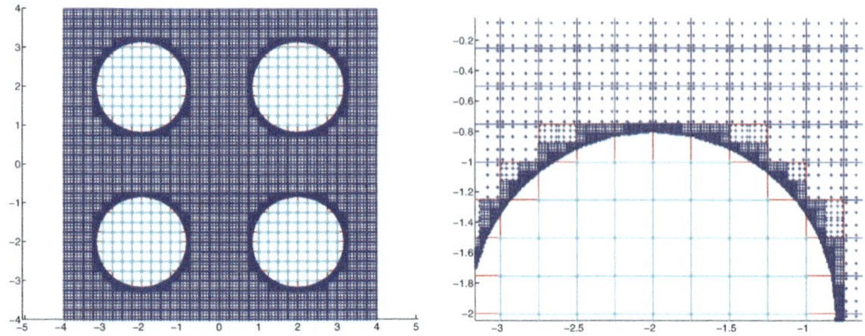

Figure 24: Numerical integration on split elements using the adaptive Gauss rule in action: Gauss points in the void space were removed.

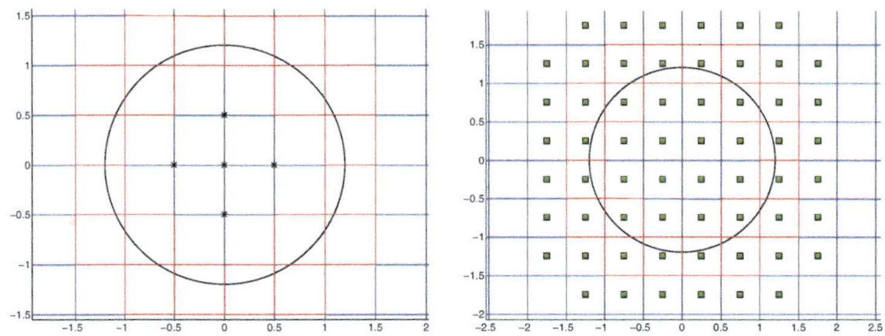

Figure 25: Nodes whose support falls within the void space are removed from the system (star nodes). Illustration for a mesh consisting of bilinear elements (left) and for a quadratic B-spline mesh with the same number of elements (right) where there are no removed dofs. Blue squares denote control points associated to split elements.

void is done with a simple criterion based on the level sets of the four nodes of the subcell under consideration. For a point $\mathbf{x}$ in an element cut by the void having $\mathbf{x}_c$ and r as the center and radius, the level set is given by

$$\phi(\mathbf{x}) = ||\mathbf{x} - \mathbf{x}_c|| - r \tag{78}$$

Note that we have not used interpolation to define the level sets at any point from the nodal level sets.

Removal of irrelevant dofs In order to avoid an ill conditioning matrix nodes whose support falls within the void space are removed from the system, we refer to Figure 25 for an example. Listing 5.2 gives an implementation of this process. In our actual implementation, we do not remove those nodes but simply fix their dofs (as zero Dirichlet BCs nodes).

Listing 5.2: Matlab code to find nodes to be removed from the system.

```
inactiveNodes1 = [];
inactiveNodes2 = [];

for ie=1:length(voidElems)
    e    = voidElems(ie);
    sctr = element(e,:);
    inactiveNodes1 = [inactiveNodes1 sctr];
end

for ie=1:length(splitElems)
    e    = splitElems(ie);
    sctr = element(e,:);
    inactiveNodes2 = [inactiveNodes2 sctr];
end

inactiveNodes1 = unique(inactiveNodes1);
inactiveNodes2 = unique(inactiveNodes2);
inactiveNodes  = setdiff(inactiveNodes1,inactiveNodes2);
```

6 Examples

In this section, numerical examples in linear elasticity and linear elastic fracture mechanics are presented with the purpose to verify the discussed Matlab code. Unless otherwise stated, standard direct imposition of Dirichlet BCs is used. The SIFs are computed using the domain form of the interaction integral and the domain is selected as the collection of elements within a circle centered at the crack tip and has a radius of $r = \alpha\sqrt{A_t}$ where α is a positive scalar and A_t denotes the area of the tip element. Unless otherwise stated, a value of 2 is chosen for α.

6.1 Two and three dimensional solid mechanics

Infinite plate with a circular hole is considered as shown in Figure 26 where, due to symmetry, only a quarter of the plate is modeled. The plate dimension is taken to be $L \times L$ and the circular hole has a radius R. The exact stress field in the plate is given by

$$\sigma_{xx}(r,\theta) = 1 - \frac{R^2}{r^2}\left(\frac{3}{2}\cos 2\theta + \cos 4\theta\right) + \frac{3}{2}\frac{R^4}{r^4}\cos 4\theta \tag{79a}$$

$$\sigma_{yy}(r,\theta) = -\frac{R^2}{r^2}\left(\frac{1}{2}\cos 2\theta - \cos 4\theta\right) - \frac{3}{2}\frac{R^4}{r^4}\cos 4\theta \tag{79b}$$

$$\sigma_{xy}(r,\theta) = -\frac{R^2}{r^2}\left(\frac{1}{2}\sin 2\theta + \sin 4\theta\right) + \frac{3}{2}\frac{R^4}{r^4}\sin 4\theta, \tag{79c}$$

where r, θ are the usual polar coordinates centered at the center of the hole.

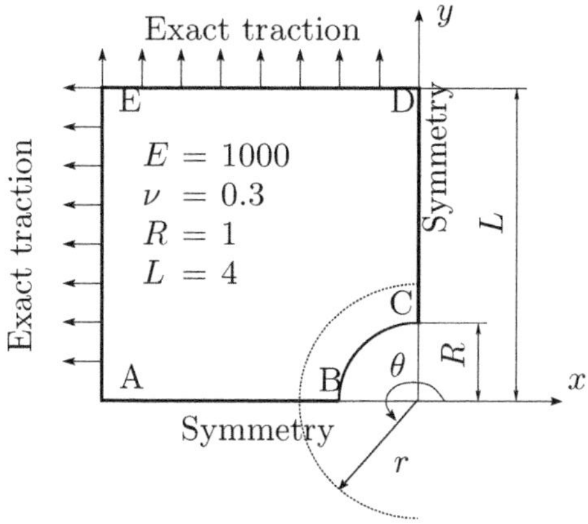

Figure 26: Infinite plate with a circular hole under constant in-plane tension: quarter model. Boundary conditions include: $u_y = 0$ (AB), $u_x = 0$ (CD), $\bar{\mathbf{t}}^{\mathrm{T}} = (-\sigma_{xx}, -\sigma_{xy})$ (AE), $\bar{\mathbf{t}}^{\mathrm{T}} = (\sigma_{xy}, \sigma_{yy})$ (ED).

The material properties are specified as $E = 10^3$, Poisson's ratio $\nu = 0.3$ and the geometry is such that $L = 4$, $R = 1$. A plane stress condition is assumed. The problem is solved with quadratic NURBS meshes such as those shown in Figure 27. The control points for the coarsest mesh can be found in Hughes et al. (2005) or file **plateHoleCkData.m**. Figure 28, generated in Paraview, illustrates the contour plot of numerical σ_{xx}. Note that the stress concentration at point $(R, 3\pi/2)$ is well captured and a smooth stress field is obtained throughout.

Note that by using the visualization technique described in Section 4.4 for this problem, a note should be made on the evaluation of the stress field at the top left corner where there are two coincident control points. This causes a singular Jacobian matrix. Therefore at this corner, the stresses at a point slightly shifted from the original position are used.

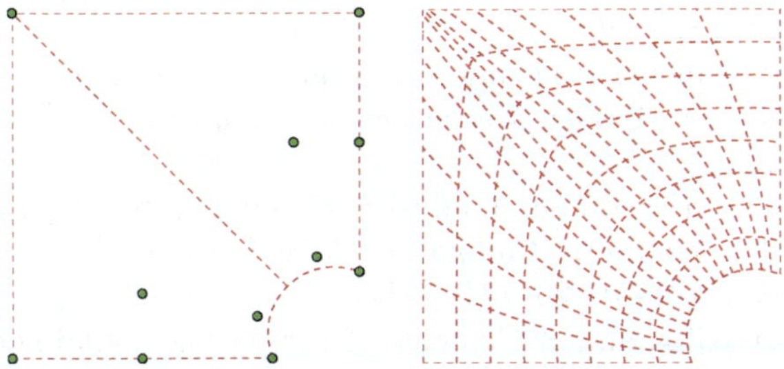

Figure 27: Plate with a hole: coarse mesh of 2 bi-quadratic elements (left) and refined mesh (right).

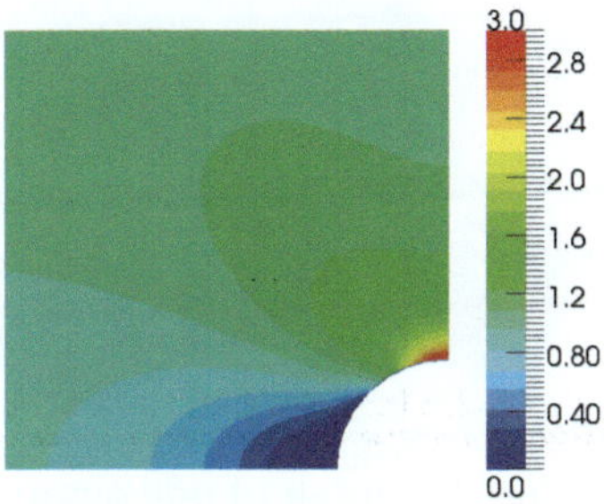

Figure 28: Plate with a hole: distribution of numerical σ_{xx} obtained with a 32×16 quadratic mesh having 4488 dofs.

Pinched cylinder In order to demonstrate the performance of the 3D IGA implementation, we consider the pinched cylinder problem as shown in Figure 29. Note that we discretize the shell with solid NURBS elements. Due to symmetry, only 1/8 of the model is analyzed. A tri-quadratic NURBS mesh ($p = q = r = 2$) was used for the computation. Details can be found in the file **igaPinchedCylinder.m**. Figure 30 shows the mesh and the contour plot of the displacement in the point load direction. Post-processing is done in Paraview and we refer to Section 4.4 for details. We recognise that the problem under consideration is a shell like structure that would be more accurately modelled using appropriate shell elements, but the example is merely intended to illustrate the ability of the method to analyse 3D geometries.

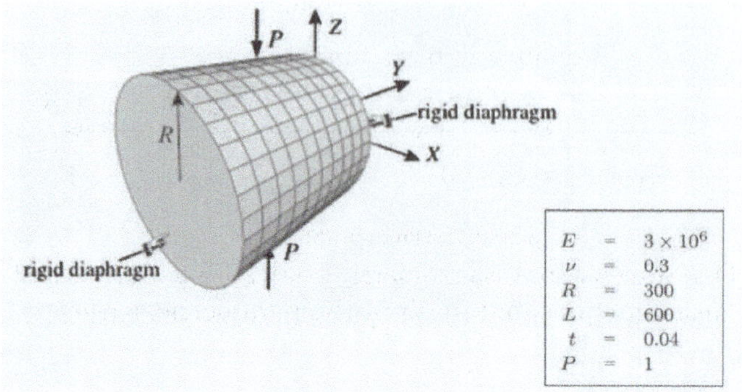

Figure 29: Pinched cylinder. Problem description and data.

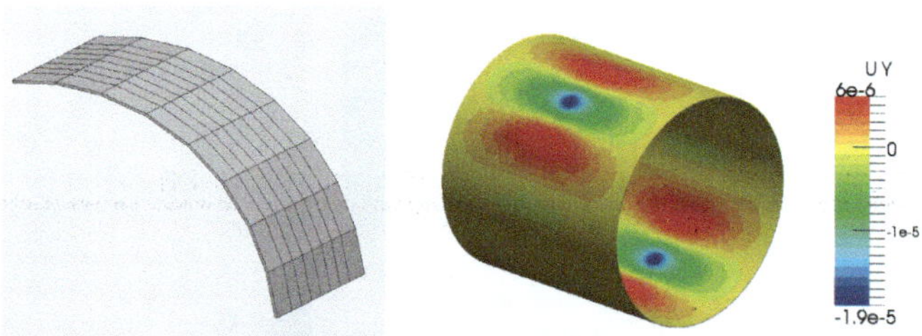

Figure 30: Pinched cylinder: (a) mesh of one octant of the cylinder and (b) contour plot of the displacement in the direction of the point load.

6.2 Incompatible multi-patch NURBS

This example concerns a multi-patch NURBS geometry and is taken from Nguyen et al. (2014c). The geometry, given in Figure 31, is a simplified representation of a connecting rod, which is a component of an internal combustion engine, and represents a classical linear case in the stress-strain static analysis. The geometric input model is composed by three NURBS patches (see Figure 32) with two incompatible patch interfaces. The dimensions are consistent with an actual component and the material properties are Young's modulus $E = 2 \times 10^5$ MPa, Poisson's ratio $\nu = 0.3$ (standard steel). Boundary conditions are represented in Figure 31: ideal fixed boundary condition on the two vertical surfaces of the (*big-end*) and a vertical total force $F = 1000$ N load applied to the internal ring of the *small-end*, according to the effect of the *pin-piston* sub-assembly that transmits a bending moment to the connecting-rod stem. For the simulation, the discretisation used consists of tri-cubic functions: $32 \times 4 \times 8$ elements for patch 1; $24 \times 12 \times 4$

elements for patch 2; and $64 \times 4 \times 8$ elements for patch 3, resulting in a total number of 4224 elements and 11305 control points. We connect the incompatible patches using a Nitsche's method Nguyen et al. (2014c). For both coupling interfaces the smaller faces are the regions where the surface integration is performed so as to ensure that all Gauss points on the elements being integrated have neighbours in the larger elements on the other patch. A stabilization parameter $\alpha = 1 \times 10^8$ was chosen empirically. The results are shown in Figure 33, where displacement and stress fields are plotted. The displacement distribution is the typical progressive cubic polynomial form of the analytical Saint-Venant model. The von Mises stress distribution is used for the comparison of the simulation results of the IGA approach with respect to *Siemens-NX* (traditional FE model, discretized with second order tetrahedra, 6182 elements and 11332 nodes Figure 34). The typical mechanical role of the connecting-rod, undergoing combined compressive and bending stress of the connecting-rod stem is visible when observing the von Mises stress distribution in the rod. These equivalent stresses are close to zero in the mean plane; the superior fibre is the locus of the maximum tensile stress, symmetrically equivalent to the compression of inferior fibres. In both analyses interesting three-dimensional effects are observed: the maximum stress values correspond to the free fibres of the stem in the superior and inferior surfaces that interact with the "big-end" of the rod; the interaction between the stem and both the big-end and small-end produces an increasing stress value in the azure region in proximity of the neutral axis that is very well described in both analysis, thus demonstrating the effectiveness of the IGA representation and of the coupling method. The boundary conditions are typically such that the system is over-determined and only the inner part of the "big-end" transmits traction/compression reactions (green regions). Due to this particular load case, parts of the "big-end" (blue regions) are superfluous in both analyses and could be suppressed, reducing the mass of the component; the internal stress distribution in the inner ring of the small-end shows again very good agreement of the combined compressive and bending stress/action behaviour that reaches the pin region.

6.3 Two and three dimensional fracture mechanics

Edge cracked plate in tension A plate of dimension $b \times 2h$ is loaded by a tensile stress $\sigma = 1.0$ psi over the top edge and bottom edge as shown in Figure 35. In the computation, the displacement along the y-axis is fixed at the bottom edge and the bottom left corner is fixed in both direction. The material parameters are $E = 10^3$ psi and $\nu = 0.3$. A plane strain condition

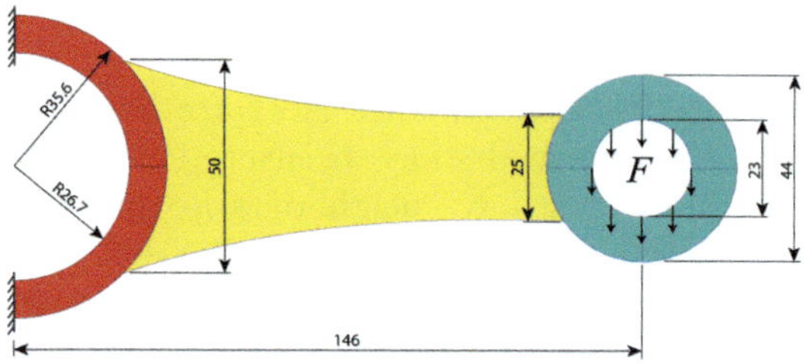

Figure 31: Connecting-rod: geometry and boundary conditions. The dimensions are in mm.

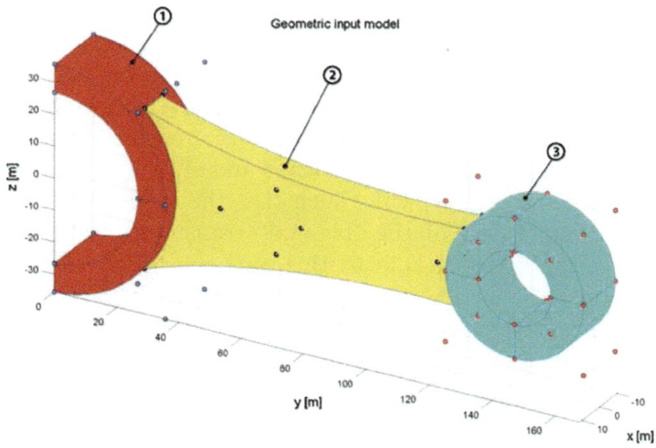

Figure 32: A multi-patch NURBS solid.

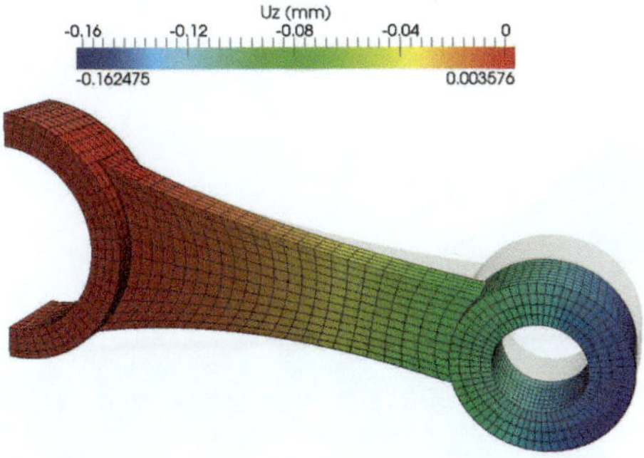

(a) z-displacement field

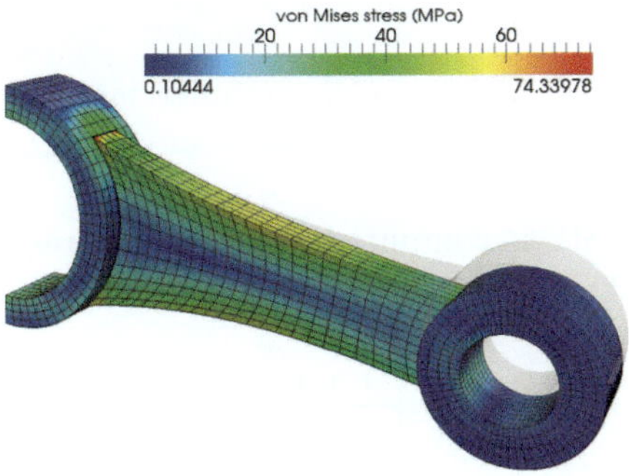

(b) Stress field

Figure 33: Results of the connecting rod.

is assumed. The reference mode I stress intensity factor (SIF) as given in Tada et al. (1985) is

$$K_I = F\left(\frac{a}{b}\right)\sigma\sqrt{\pi a} \tag{80}$$

where a is the crack length, b is the plate width, and $F(a/b)$ is an empirical function. For $a/b \leq 0.6$, the function F is given by

$$F\left(\frac{a}{b}\right) = 1.12 - 0.23\left(\frac{a}{b}\right) + 10.55\left(\frac{a}{b}\right)^2 - 21.72\left(\frac{a}{b}\right)^3 + 30.39\left(\frac{a}{b}\right)^4 \tag{81}$$

We are solving this problem using both XFEM and extended isogeometric

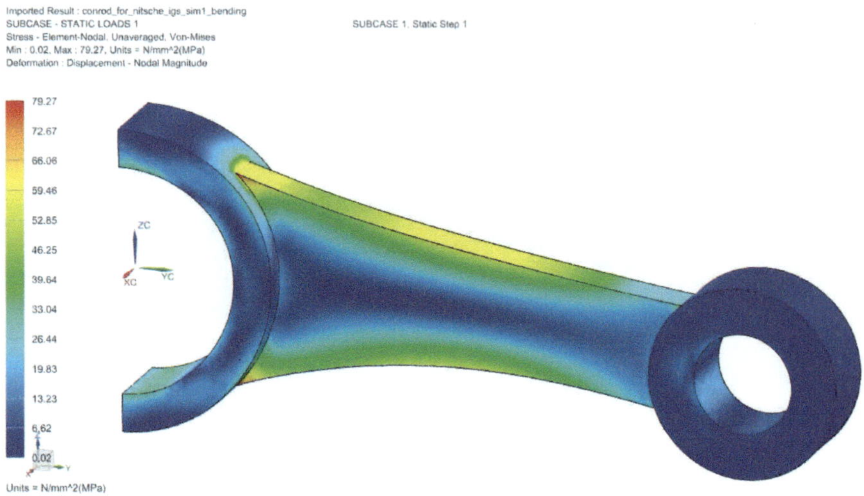

Figure 34: Stress plot from the commercial code *NX-NASTRAN*.

formulation. The SIF is computed using the interaction integral. We refer to Moës et al. (1999) for details.

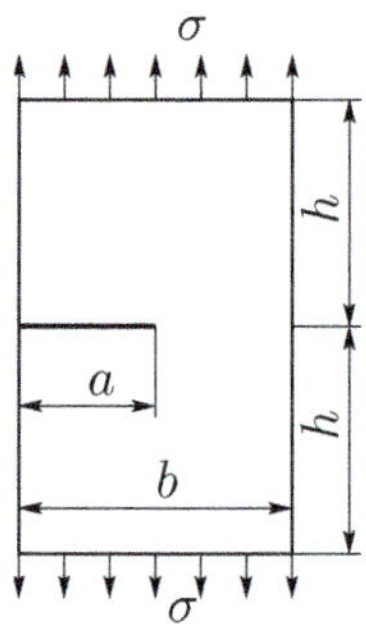

Figure 35: Edge cracked plate in tension: geometry and loading.

We first check the implementation of the XIGA code by comparing the XIGA result with the XFEM result for the case of $a = 0.45$, $b = 1$ and $h = 1$. The XFEM and XIGA meshes are given in Figure 36. Both meshes have the same uniform distribution of nodes and in the XIGA, cubic ($p = q = 3$) B-spline basis functions are adopted. The input file of this example is given in List 6.1. For this simple geometry, the control points are simply the nodes of a structured mesh (line 6). The main script M file for this problem is **xigaEdgeCrack.m**. Figure 37 shows the contour plots of the vertical displacement obtained with XFEM and XIGA.

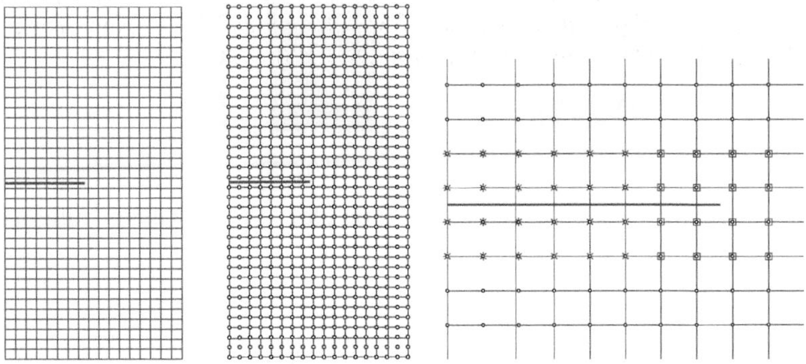

(a) XFEM mesh (Q4) (b) XIGA mesh (cubic Bsplines) (c) XIGA mesh, enriched nodes

Figure 36: Edge cracked plate: XFEM and XIGA meshes. Both have the same number of displacement dofs of 1296. The thick line denotes the crack.

Listing 6.1: File edgeCrackC2Data.m for the edged crack problem.

```
p = 3; q = 3;
L = 1;
noPtsX = 56;
noPtsY = 112;

gcoord=meshRectangularCoord(1,2,noPtsX-1,noPtsY-1);
controlPts=gcoord;

weights = ones(1,noPtsX*noPtsY)';

knotUTemp = linspace(0,1,noPtsX-p+1);
knotVTemp = linspace(0,1,noPtsY-q+1);

uKnot     = [0 0 0 knotUTemp 1 1 1];
vKnot     = [0 0 0 knotVTemp 1 1 1];

noCtrPts  = noPtsX * noPtsY;
noDofs    = noCtrPts * 2;
% generate element connectivity ...
generateIGA2DMesh
```

We now move to the computation of the mode I SIF. We consider a crack of length $a = 0.3$. The reference SIF is $K_I^{\text{ref}} = 1.6118$. Both linear

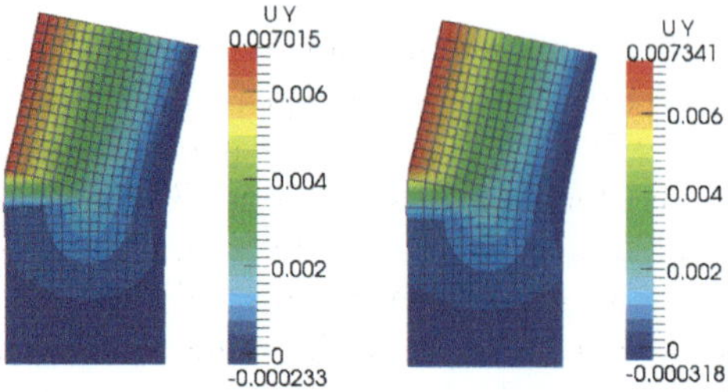

(a) XFEM (b) XIGA

Figure 37: Edge cracked plate: u_y contour plots on deformed configuration (factor of 30).

and cubic B-spline basis are used with three meshes. The result is given in Table 1. Excellent agreement with the analytical solution was obtained. It should be emphasized that, in addition to the standard enrichment scheme, the geometrical enrichment scheme in which a layer of elements around the tip elements are also enriched, see Figur 38. This option is not the default one, for a typical usage, we refer to file **edgeCrackC2Data.m**.

mesh	disp. dofs	K_I (linear)	Error (%)	K_I (cubic)	Error (%)
9×18	324	1.4997	6.96	1.5560	3.46
18×36	1296	1.5823	1.83	1.6150	0.20
36×72	5184	1.5968	0.93	1.6117	0.01

Table 1: Edge cracked plate: SIFs results. The reference SIF is $K_I^{\text{ref}} = 1.6118$.

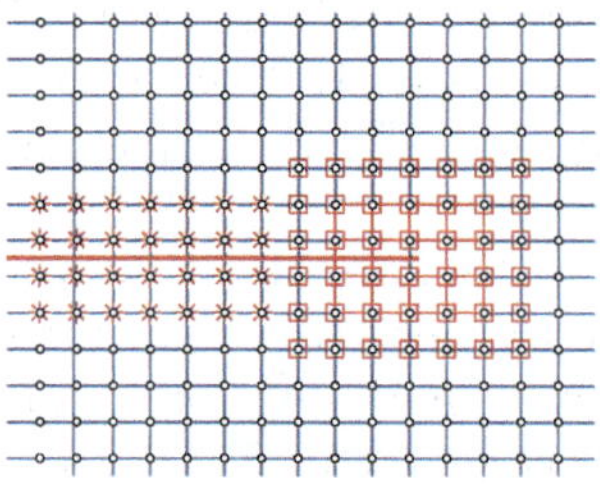

Figure 38: Edge cracked plate: geometrical enrichment on a cubic mesh.

Infinite plate with a center crack Consider an infinite plate containing a straight crack of length $2a$ and loaded by a remote uniform stress field σ. Along ABCD the closed form mode I displacement field in terms of polar coordinates in a reference frame (r, θ) centered at the crack tip is:

$$
\begin{aligned}
u_x(r,\theta) &= \frac{K_I}{2\mu}\sqrt{\frac{r}{2\pi}}\cos\frac{\theta}{2}\left(\kappa - 1 + 2\sin^2\frac{\theta}{2}\right)\\
u_y(r,\theta) &= \frac{K_I}{2\mu}\sqrt{\frac{r}{2\pi}}\sin\frac{\theta}{2}\left(\kappa + 1 - 2\cos^2\frac{\theta}{2}\right)
\end{aligned}
\tag{82}
$$

where $\mu = \frac{E}{2(1+\nu)}$ is the shear modulus and κ is the Kolosov constant that is given by

$$
\kappa = \begin{cases} 3-4\nu & \text{plane strain}\\ \dfrac{3-\nu}{1+\nu} & \text{plane stress} \end{cases}
\tag{83}
$$

and the mode I stress field is given by

$$
\begin{aligned}
\sigma_{xx}(r,\theta) &= \frac{K_I}{\sqrt{2\pi r}}\cos\frac{\theta}{2}\left(1 - \sin\frac{\theta}{2}\sin\frac{3\theta}{2}\right)\\
\sigma_{yy}(r,\theta) &= \frac{K_I}{\sqrt{2\pi r}}\cos\frac{\theta}{2}\left(1 + \sin\frac{\theta}{2}\sin\frac{3\theta}{2}\right)\\
\sigma_{xy}(r,\theta) &= \frac{K_I}{\sqrt{2\pi r}}\sin\frac{\theta}{2}\cos\frac{\theta}{2}\cos\frac{3\theta}{2}
\end{aligned}
\tag{84}
$$

where $K_I = \sigma\sqrt{\pi a}$ is the mode I stress intensity factor, ν is Poisson's ratio and E is Young's modulus. ABCD is a square of $10 \times 10\text{mm}^2$, $a = 100\text{mm}$; $E = 10^7\text{N/mm}^2$, $\nu = 0.3$, $\sigma = 10^4\text{N/mm}^2$.

The geometry of the computational domain ABCD is shown in Figure 39. Displacement of nodes on the bottom, right and top edges are prescribed by Equation (86) while along the left edge a traction defined by Equation (87) is applied. Note that on the left edge there are additional dofs due to enrichment, the enforcement of essential BCs on this edge is thus complicated. We decided to impose a traction there instead.

The problem is solved with a cubic mesh (the M file is **xigaInfiniteCrackModeI.m** [6].) and the comparison between the numerical deformed

[6] There exists also file **xigaInfiniteCrackModeILM.m** (Lagrange multipliers are used to enforce the Dirichlet BCs) and **xigaInfiniteCrackModeILeastSquare.m** (Least Square method is used for Dirichlet BCs)

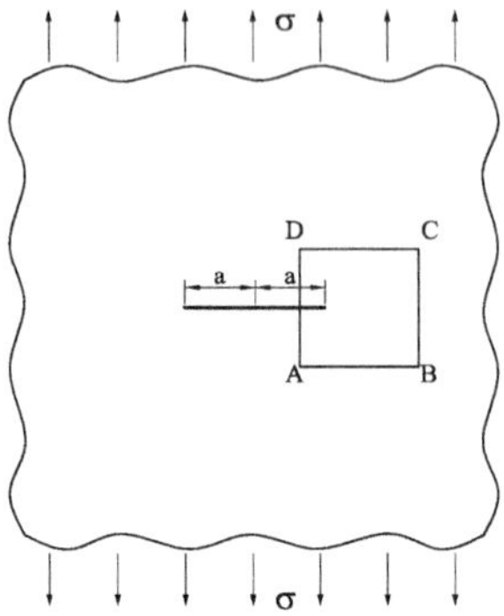

Figure 39: Infinite cracked plate under remote tension: geometry and loads

configuration and exact deformed configuration is given in Figure 40. Figure 41 shows the contour plot of the von Mises equivalent stresses σ_{eqv} obtained with the simulation and the analytical solution. The von Mises stress is defined as

$$\sigma_{eqv} = \sqrt{\sigma_{xx}^2 + \sigma_{yy}^2 - \sigma_{xx}\sigma_{yy} + 3\sigma_{xy}^2} \tag{85}$$

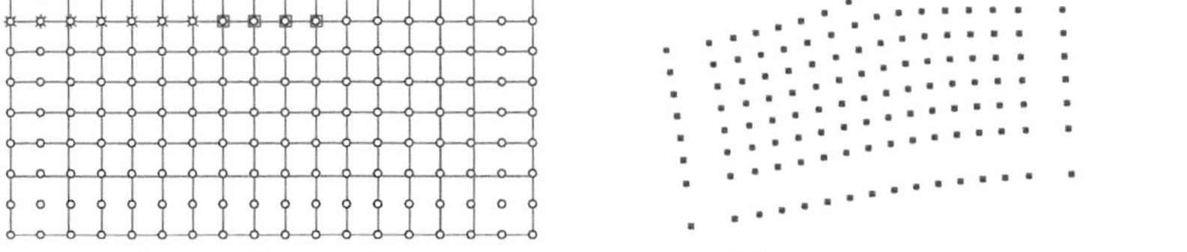

(a) enriched nodes (b) deformed shape

Figure 40: Infinite plate with a crack under mode I loading: enrichment and comparison between the numerical and exact displacements. A cubic mesh of 15 × 15 elements is used.

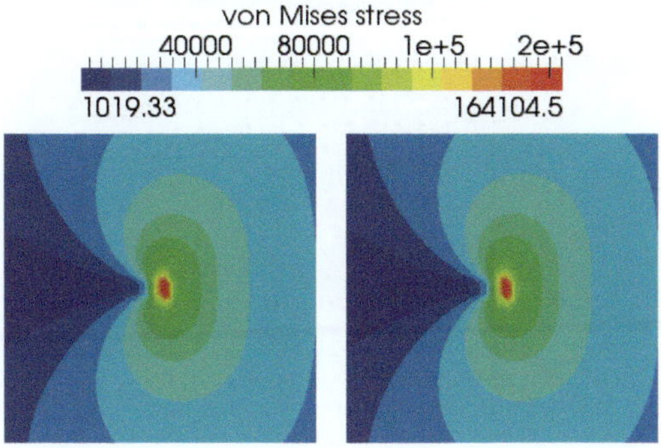

Figure 41: Infinite cracked plate under remote tension: von Mises equivalent stress contour obtained with IGA (left) and exact solution (right). A cubic mesh of 33 × 33 elements is used.

We now turn our attention to the mode II problem in which the plate is subjected to a remote shear stress. The mode II closed form displacements are given by

$$\begin{aligned} u_x(r,\theta) &= \frac{K_{II}}{2\mu}\sqrt{\frac{r}{2\pi}}\sin\frac{\theta}{2}\left(\kappa+1+2\cos^2\frac{\theta}{2}\right) \\ u_y(r,\theta) &= \frac{K_{II}}{2\mu}\sqrt{\frac{r}{2\pi}}\cos\frac{\theta}{2}\left(1-\kappa+2\sin^2\frac{\theta}{2}\right) \end{aligned} \tag{86}$$

where $K_{II} = \sigma\sqrt{\pi a}$ is the mode II stress intensity factor, and the mode II stress field is given by

$$\begin{aligned} \sigma_{xx}(r,\theta) &= -\frac{K_{II}}{\sqrt{2\pi r}}\sin\frac{\theta}{2}\left(2+\cos\frac{\theta}{2}\cos\frac{3\theta}{2}\right) \\ \sigma_{yy}(r,\theta) &= \frac{K_{II}}{\sqrt{2\pi r}}\sin\frac{\theta}{2}\cos\frac{\theta}{2}\cos\frac{3\theta}{2} \\ \sigma_{xy}(r,\theta) &= \frac{K_{II}}{\sqrt{2\pi r}}\cos\frac{\theta}{2}\left(1-\sin\frac{\theta}{2}\sin\frac{3\theta}{2}\right) \end{aligned} \tag{87}$$

The M file is **xigaInfiniteCrackModeII.m** and the results are given in Figures 42 and 43.

Edge crack under shear stress A plate is clamped on the bottom edge and loaded by a shear traction $\tau = 1.0$ psi over the top edge as shown in

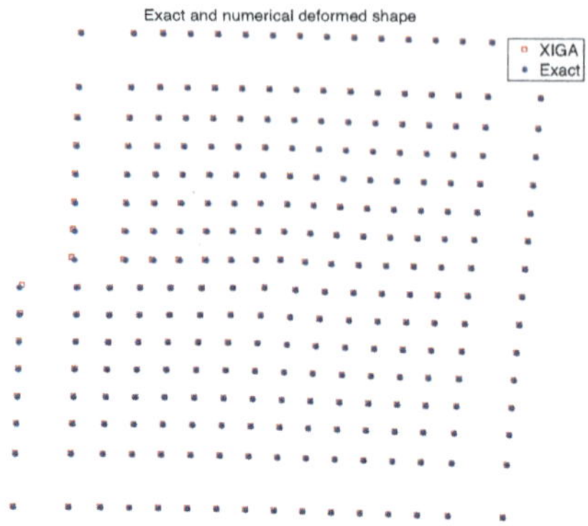

Figure 42: Infinite plate with a crack under mode II loading: comparison between the numerical and exact displacements. A cubic mesh of 15×15 elements is used.

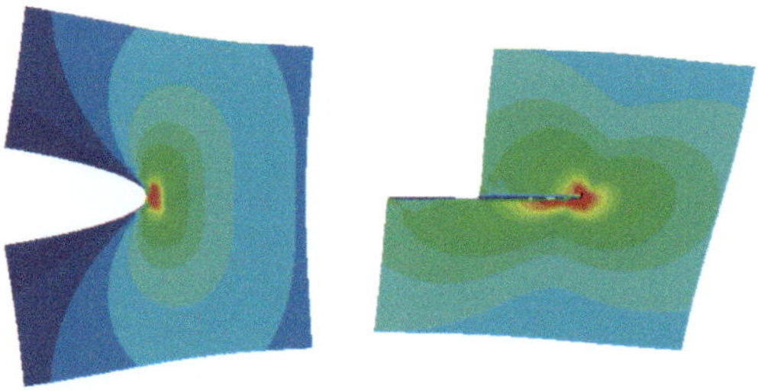

Figure 43: Infinite cracked plate: von Mises equivalent stress contours on cracked meshes.

Figure 44. The material parameters are $3 \times 10^7 psi$ for Young's modulus and 0.25 for Poisson's ratio. The reference mixed mode stress intensity factors are

$$\begin{aligned} K_I &= 34.0 \quad psi\sqrt{in} \\ K_{II} &= 4.55 \quad psi\sqrt{in} \end{aligned} \tag{88}$$

The M file for this problem is **xigaEdgeShearCrack.m** and the data file is **edgeShearCrackCkData.m**.

In what follows we present a technique to create mesh with a certain level of refinement at regions of interest such as around cracks. The basic idea is

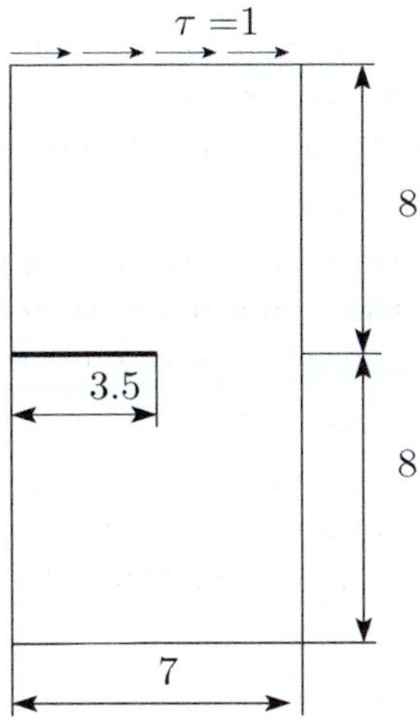

Figure 44: Geometry and load of the shear edge crack problem.

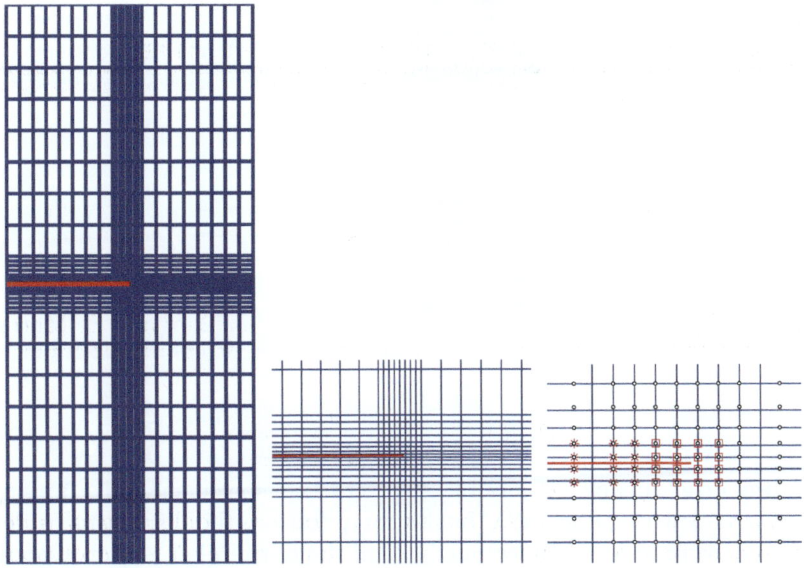

Figure 45: Edge crack under shear: mesh of 24 × 31 cubic elements with global refinement.

to use the *linear parameterization* and it is based on the property of NURBS that order elevation does not change the NURBS parametrically. Specially, we first build a linear NURBS surface with knot vectors $\Xi = \{0, 0, 1, 1\}$ and $\mathcal{H} = \{0, 0, 1, 1\}$. There are four control points. Then, the surface is order elevated to a higher order in both directions. It should be emphasized that the parameterization of this high order surface is still linear. In the next step, h-refinement is adopted. The knot values $\{0.43, 0.55\}$ are inserted in the ξ direction and $\{0.45, 0.55\}$ are inserted in the η direction. This constitutes the coarse mesh from which knot insertion by bisecting the knot spans is performed to obtain more refined meshes. It should be emphasized that in order for the crack not align the element edge, the knot span (vertical direction) contains the crack is divided into three sub-spans rather than two as usual. One such mesh is given in Figure 45. This example also demonstrated the deficiency of the tensor product based NURBS–they do not support local refinement.

Figure 46 gives a contour of the von Mises stress. The computed SIFs are $K_I = 33.96\, psi\sqrt{in}$ and $K_{II} = 4.92\, psi\sqrt{in}$.

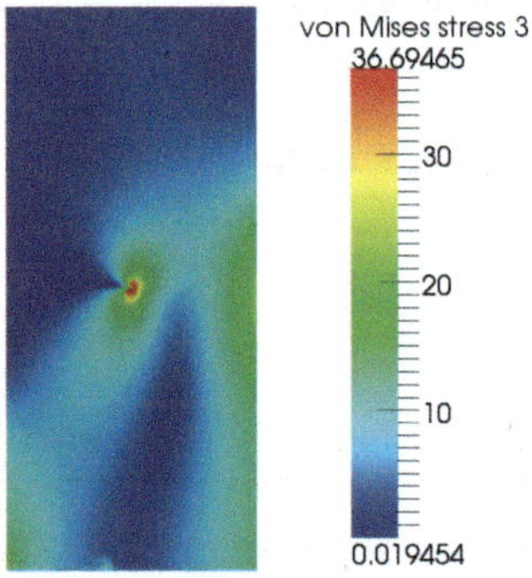

Figure 46: Edge crack under shear: von Mises stress distribution.

Double edge crack A double edge cracked specimen under uni-axial tension is shown in Fig. (47). In the computation, the displacement along the y-axis is fixed at the bottom edge and the bottom left corner is fixed in both direction. A tensile stress of $\sigma = 1.0$ psi is applied on the top edge. The material parameters are $E = 10^3$ psi and $\nu = 0.3$. A plane strain condition is assumed. Geometry data are $b = 1$ and $h = 1$.

The mode I SIF is given by

$$K_I = F\left(\frac{a}{b}\right)\sigma\sqrt{\pi a} \tag{89}$$

where

$$F\left(\frac{a}{b}\right) = 1.12 + 0.43\left(\frac{a}{b}\right) - 4.79\left(\frac{a}{b}\right)^2 + 15.46\left(\frac{a}{b}\right)^3 \tag{90}$$

One goal of this example is to illustrate the implementation of multiple cracks in an XIGA framework. For this problem we need to know which nodes are enriched by which cracks. We solve this problem using a quartic ($p = q = 4$) mesh as shown in Figure 48. Note that the mesh is created from a uniform distribution of control points. The mesh in the physical space is however not uniform. This is because the basis functions near the beginning and end of the domain are not identical to those in the interior (recall Figure 1).

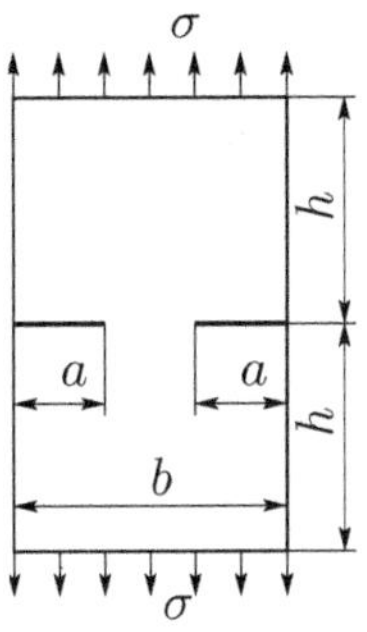

Figure 47: Double edge cracked plate in tension: geometry and loading.

The M file for this problem is **xigaTwoEdgeCracks.m.** Figure 49 shows the plot of the vertical displacement field and the von Mises stress.

Center crack in an finite plate under tension Consider a 2×2 square ($W = 2$) that contains a centered crack of length $2a = 0.5$. The plate is under tension in the vertical direction of magnitude $\sigma = 1$. The Young's modulus is 3×10^7 and Poison's ratio is 0.25. The analytical SIF is

$$K_I = \sqrt{\sec\frac{\pi a}{W}}\sigma\sqrt{\pi a} \tag{91}$$

For the chosen parameters, the mode I SIF is 0.922.

The goal of this example is to show how cracks with two tips are modelled in MIGFEM. The M file for this problem is **xigaCenterCrack.m.** Figure 50 shows the adopted mesh and the stress contour. The computed SIFs at two tips are 0.944.

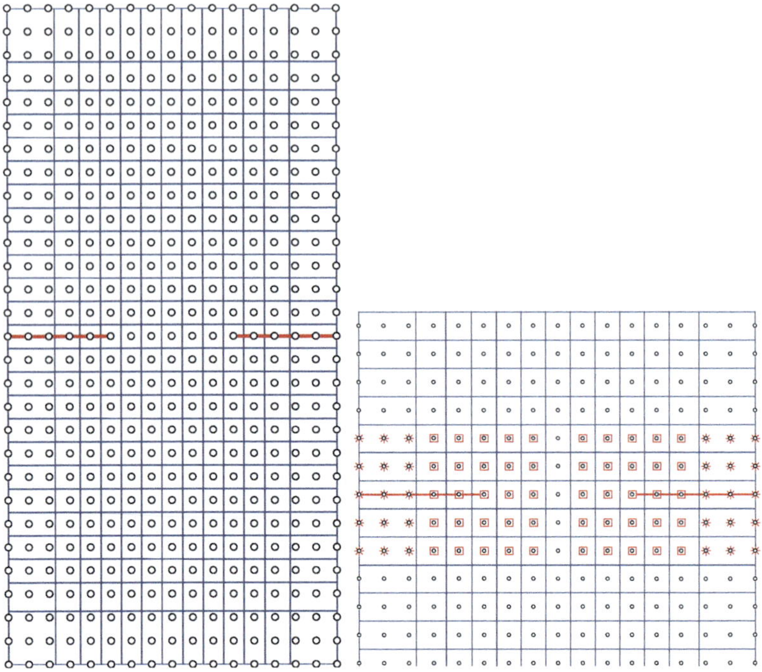

(a) mesh (b) enriched nodes

Figure 48: Double edge cracked plate in tension: a quartic ($p = q = 4$) mesh and enriched nodes.

6.4 Three-dimensional mode I fracture problem

This example aims to show the capability of our Matlab code for solving three-dimensional (3D) fracture problems. For 3D cracks, the polar coordinates in the branch functions given in Equation (74) are defined in terms of the level sets as Shi et al. (2010)

$$r = \sqrt{\varphi(\xi,\eta,\zeta)^2 + \psi(\xi,\eta,\zeta)^2}, \quad \theta = \operatorname{atan}\left(\frac{\varphi(\xi,\eta,\zeta)}{\psi(\xi,\eta,\zeta)}\right) \tag{92}$$

where the level set field $\Phi = (\varphi, \psi)$ are interpolated as

$$\Phi(\xi,\eta,\zeta) = \sum_I \phi_I(\xi,\eta,\zeta)\Phi_I \tag{93}$$

We refer to Shi et al. (2010) for details concerning the derivatives of the branch functions with respect to the parametric coordinates (ξ, η, ζ).

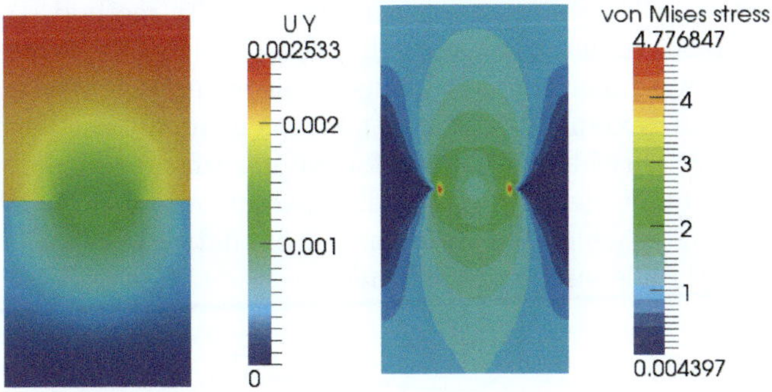

Figure 49: Double edge cracked plate in tension: contour plots of the vertical displacement and the von Mises stress.

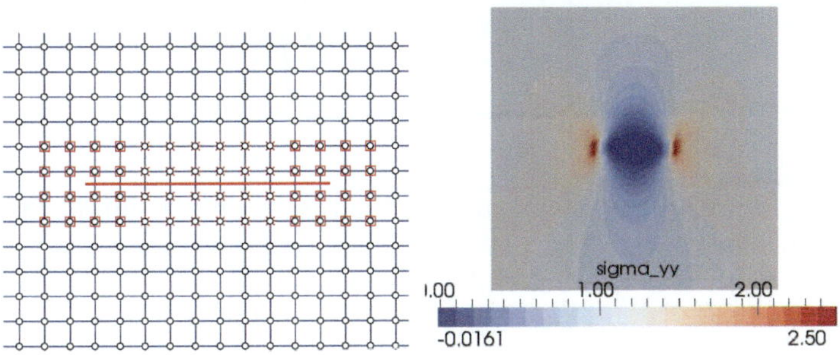

Figure 50: Center cracked plate in tension: mesh of 37×37 cubic elements (left) and σ_{yy} contour (right).

The mode I 3D fracture problem we are solving is given in Figure 51. The exact displacement field is given by

$$
\begin{aligned}
u_x(r,\theta) &= \frac{2(1+\upsilon)}{\sqrt{2\pi}}\frac{K_I}{E}\sqrt{r}\cos\frac{\theta}{2}\left(2-2\upsilon-\cos^2\frac{\theta}{2}\right)\\
u_y(r,\theta) &= 0 \qquad\qquad (94)\\
u_z(r,\theta) &= \frac{2(1+\upsilon)}{\sqrt{2\pi}}\frac{K_I}{E}\sqrt{r}\sin\frac{\theta}{2}\left(2-2\upsilon-\cos^2\frac{\theta}{2}\right)
\end{aligned}
$$

where $K_I = \sigma\sqrt{\pi a}$ is the stress intensity factor, υ is Poisson's ratio and E is Young's modulus. The data are $a = 100$ mm; $E = 10^7$ N/mm^2, $\upsilon = 0.3$, $\sigma = 10^4$ N/mm^2. On the bottom, right and top surfaces, essential BCs taken from Equation (94) are imposed using the penalty method. A penalty parameter of $1e10$ was used. We note that this problem can be more effectively solved with two-dimensional elements. This example however aims at presenting how 3D extended IGA can be implemented. Furthermore, it also illustrates how Dirichlet BCs are enforced on surfaces rather than the usual case of line boundaries. To this end, a two-dimensional NURBS mesh for a given surface is generated from the set of control points that define this surface (see the file **surfaceMesh.m**).

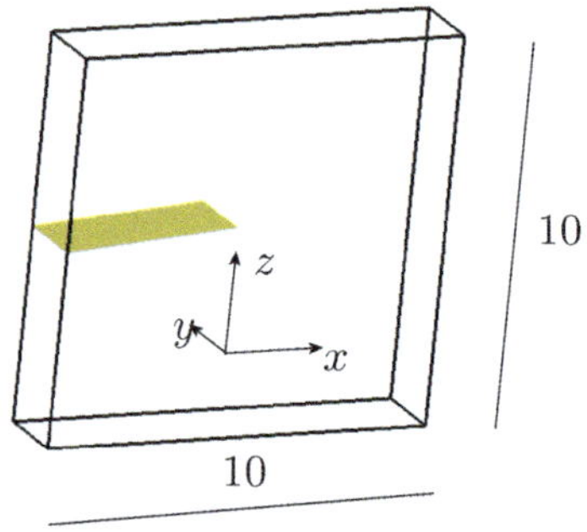

Figure 51: Three-dimensional mode I fracture problem: infinite plate with a center planar crack. The plate thickness is 2, the crack length is 5 and the crack width is 2. The crack locates in the mid-plane of the plate.

The problem is firstly solved with a linear NURBS basis. A mesh of $9 \times 9 \times 1$ elements are used. The mesh, enriched nodes and comparison of the numerical deformed configuration against the exact one are given in Figure 52. Next, a mesh of $7 \times 7 \times 2$ elements where along the thickness there are two linear elements ($q = 1$) and for the other directions, quadratic basis ($p = r = 2$) are adopted. The result is given in Figure 53. The M file is **xigaInfiniteCrack3d.m**.

7 Structural mechanics

Due to the inherent high order continuity that B-splines/NURBS/T-splines possess implementation of models that require C^k ($k \geq 1$) finite elements is straightforward. Such models include Euler-Bernoulli beams, Kirchhoff plates and Kirchhoff-Love shells models as far as structural mechanics is concerned. Other high order models include gradient elasticity Fischer et al. (2010) and gradient enhanced damage models Verhoosel et al. (2011b). In this section, the formulation of isogeometric finite element formulations for

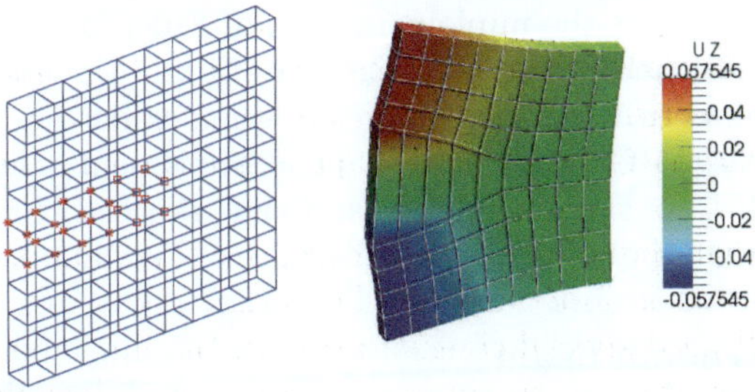

Figure 52: Three-dimensional mode I fracture problem: mesh of linear NURBS elements and enriched control points (left); numerical deformed configuration (magnification factor of 40) superimposed on the exact deformed configuration (right).

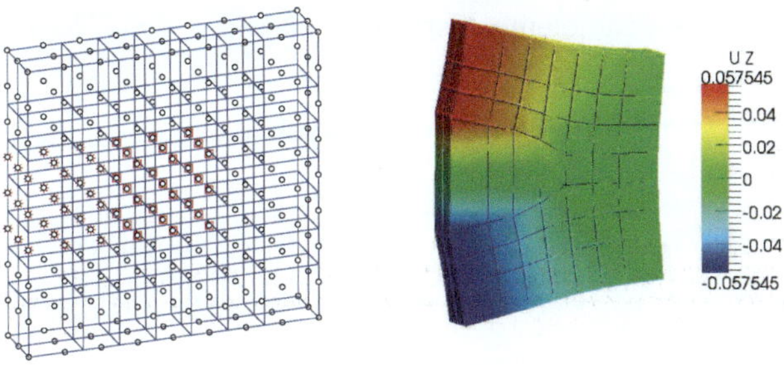

Figure 53: Three-dimensional mode I fracture problem: mesh of quadratic NURBS elements and enriched control points (left); numerical deformed configuration (magnification factor of 40) superimposed on the exact deformed configuration (right).

thin beams, plates and shells is given. Only expressions for the stiffness matrix of isogeometric beam and plate elements are provided as the theory of beams/plates is classic and can be found for instance in Reddy (2005).

The commonly used plate models are listed as follows

- Kirchhoff model: for thin bent plates with small deflections, negligible shear energy and uncoupled membrane-bending action;
- Reissner-Mindlin model: for thin and moderately thick bent plates in which first-order transverse shear effects are considered;
- Von-Karman model: for very thin bent plates.

The Kirchhoff plate model requires the use of C^1 finite elements.

This section presents the implementation of Euler-Bernoulii beam, Timoshenko beam, Kirchhoff plate and Reissner-Mindlin plate using NURBS. The code described in this section can be found under folder **structural-mechanics** of the MIGFEM package. The implementation of the isogeometric Kirchhoff-Love shell is lengthy and thus skipped. Interested readers can refer to Kiendl et al. (2009) for details.

7.1 Euler-Bernoulli beam

Euler-Bernoulli beam finite elements have often been constructed using the Hermite cubic polynomials that involve both transverse and rotation degrees of freedom. By using B-splines/NURBS of order equal or greater than two as the finite element basis the C^1 continuity condition can be easily satisfied without resorting to rotation dofs. We refer to file **igaBeam1D.m** for the implementation.

For completeness the element stiffness matrix is given here

$$\mathbf{K}^e = \int_\Omega \mathbf{B}_e EI \mathbf{B}_e^{\mathrm{T}} \mathrm{d}x \tag{95}$$

in which $I = bh^3/12$ and $\mathbf{B}_e$ denotes a vector containing the second derivatives of the shape functions with respect to x. The second derivatives of a shape function with respect to the global coordinates will be discussed in section 7.1.

Boundary conditions For a rotation-free beam formulation, special treatment of clamped boundary conditions should be taken. As can be seen from Figure 54, the tangent of a B-spline/NURBS curve at control point 1 is determined by this point and the one next to it (control point 2). Therefore, in order to model a clamped BC at point 1, one only needs to fix the displacements of both points 1 and 2.

Second derivatives of shape functions One firstly computes the second derivative of a shape function with respect to the natural coordinate as follows

$$\frac{\partial^2 f}{\partial \xi^2} = \frac{\partial}{\partial \xi}\left(\frac{\partial f}{\partial \xi}\right) = \frac{\partial}{\partial \xi}\frac{\partial f}{\partial x}\frac{\partial x}{\partial \xi} = f_x x_{\xi\xi} + x_\xi^2 f_{xx} \tag{96}$$

where the subscripts are used to indicate differentiation. Writing this for all the shape functions of an element e and put them in a matrix form, one

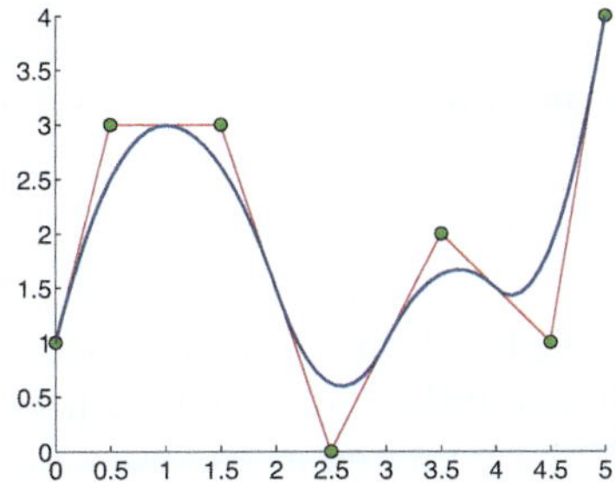

Figure 54: A quadratic B-spline curve with a uniform open knot vector $\Xi = \{0, 0, 0, 1, 2, 3, 4, 5, 5, 5\}$. Control points are denoted by filled circles.

can write (assuming, for clarity, two shape functions)

$$\begin{bmatrix} f^1_{xx} \\ f^2_{xx} \end{bmatrix} = \frac{1}{x^2_\xi}\left(\begin{bmatrix} f^1_{\xi\xi} \\ f^2_{\xi\xi} \end{bmatrix} - x_{\xi\xi}\begin{bmatrix} f^1_x \\ f^2_x \end{bmatrix}\right) \tag{97}$$

7.2 Timoshenko beam

The transverse displacement and the rotation are approximated using the same shape functions as follows

$$w = N_I w_I, \quad \beta = N_I \beta_I \tag{98}$$

where (w_I, β_I) denote the nodal dofs (deflection and rotation); N_I can be any B-spline/NURBS/T-spline basis function.

For a two noded linear element, the element nodal unknowns are written as $\mathbf{a}_e = [w_1, \beta_1, w_2, \beta_2]$. Therefore, one can write

$$\beta_{,1} = \underbrace{\begin{bmatrix} 0 & N_{1,1} & 0 & N_{2,1} \end{bmatrix}}_{\mathbf{B}_{be}} \begin{bmatrix} w_1 \\ \beta_1 \\ w_2 \\ \beta_2 \end{bmatrix}, \quad \beta + w_{,1} = \underbrace{\begin{bmatrix} N_{1,1} & N_1 & N_{2,1} & N_2 \end{bmatrix}}_{\mathbf{B}_{se}} \begin{bmatrix} w_1 \\ \beta_1 \\ w_2 \\ \beta_2 \end{bmatrix} \tag{99}$$

And the element stiffness matrix then reads

$$\mathbf{K}^e = \frac{Ebh^3}{12}\int_{\Omega_e} \mathbf{B}^{\mathrm{T}}_{be}\mathbf{B}_{be} dx + kGbh \int_{\Omega_e} \mathbf{B}^{\mathrm{T}}_{se}\mathbf{B}_{se} dx \tag{100}$$

in which k is the shear factor which usually takes a value of $5/6$ and G denotes the shear modulus which equals $E/2/(1+\nu)$. The shear correction factor k necessary because across thickness shear stresses are parabolic according to elasticity theory but constant according to Timoshenko beam theory. We refer to file **igaTimoshenkoBeam1D.m** for the implementation.

7.3 Kirchhoff plate

The extension of the Euler-Bernoulli beam theory to plates is the Kirchhoff plate theory which is suitable only for thin plates.

The deflection w is approximated as follows

$$w = N_I(\xi, \eta) w_I \tag{101}$$

where $N_I(\xi, \eta)$ is the B-splines/NURBS/T-splines basis function associated with node I and w_I is the nodal deflection.

The element stiffness matrix is defined as

$$\mathbf{K}_e = \int_{\Omega_e} \mathbf{B}_e^{\mathrm{T}} \mathbf{D} \mathbf{B}_e \mathrm{d}\Omega \tag{102}$$

where the constitutive matrix $\mathbf{D}$ reads

$$\mathbf{D} = \frac{Eh^3}{12(1-\nu^2)} \begin{bmatrix} 1 & \nu & 0 \\ \nu & 1 & 0 \\ 0 & 0 & 0.5(1-\nu) \end{bmatrix} \tag{103}$$

and the element displacement-curvature matrix $\mathbf{B}$ is given by

$$\mathbf{B}_e = \begin{bmatrix} R_{1,xx} & R_{2,xx} & \cdots R_{n,xx} \\ R_{1,yy} & R_{2,yy} & \cdots R_{n,yy} \\ 2R_{1,xy} & 2R_{2,xy} & \cdots 2R_{n,xy} \end{bmatrix} \tag{104}$$

where n denotes the number of basis functions of element e. The second derivatives of bivariate shape functions with respect to the physical coordinates are described in Section 7.3.

Mass matrix For transient analysis and modal analysis one needs the mass matrix. We start with the virtual kinetic work given by

$$\begin{aligned} \delta W^{\mathrm{kin}} &= \int_V \rho \delta u_i \dot{u}_i \mathrm{d}V = \int_V \rho (x_3^2 \delta w_{,1} \dot{w}_{,1} + x_3^2 \delta w_{,2} \dot{w}_{,2} + \delta w \dot{w}) \mathrm{d}V \\ &= \int_\Omega \rho \left[\frac{t^3}{12} (\delta w_{,1} \dot{w}_{,1} + \delta w_{,2} \dot{w}_{,2}) + t \delta w \dot{w} \right] \mathrm{d}\Omega \end{aligned} \tag{105}$$

where in the second equality, through-the-thickness integration has been carried out. Introducing the finite element approximation of the transverse displacement w yields

$$\delta W^{\mathrm{kin}} = \int_\Omega \rho \left[\frac{t^3}{12}(N_{I,1}\delta w_I \ddot{w}_J N_{J,1} + N_{I,2}\delta w_I N_{J,2}\ddot{w}_J) + tN_I \delta w_I N_J \ddot{w}_J \right] \mathrm{d}\Omega \tag{106}$$

Therefore, the mass matrix is given by

$$M_{IJ} = \int_\Omega \left[\frac{\rho t^3}{12}(N_{I,1}N_{J,1} + N_{I,2}N_{J,2}) + \rho t N_I N_J \right] \mathrm{d}\Omega \tag{107}$$

And the element mass matrix is then written as

$$M_e = \int_\Omega \left[\frac{\rho t^3}{12}(\mathbf{dN}_e)^{\mathrm{T}}\mathbf{dN}_e + \rho t(\mathbf{N}_e)^{\mathrm{T}}\mathbf{N}_e \right] \mathrm{d}\Omega \tag{108}$$

where $\mathbf{dN}_e$ is a matrix containing the first derivatives of the shape functions

$$\mathbf{dN}_e = \begin{bmatrix} N_{1,x} & N_{2,x} & \cdots & N_{n,x} \\ N_{1,y} & N_{2,y} & \cdots & N_{n,y} \end{bmatrix} \tag{109}$$

in which n denotes the number of nodes/control points of the element and $\mathbf{N}_e$ denotes the row vector containing the shape functions.

Second derivatives of shape functions The derivatives of a shape function f with respect to the natural coordinate system are given by

$$\begin{aligned} f_\xi &= f_x x_\xi + f_y y_\xi \\ f_\eta &= f_x x_\eta + f_y y_\eta \end{aligned} \tag{110}$$

or in matrix form as follows

$$\begin{bmatrix} f_\xi \\ f_\eta \end{bmatrix} = \begin{bmatrix} x_\xi & y_\xi \\ x_\eta & y_\eta \end{bmatrix} \begin{bmatrix} f_x \\ f_y \end{bmatrix} \tag{111}$$

in which the 2×2 matrix is the Jacobian matrix $\mathbf{J}$. Let us assume that our element has three nodes, hence three shape functions. We can write

$$\begin{bmatrix} f_x^1 & f_x^2 & f_x^3 \\ f_y^1 & f_y^2 & f_y^3 \end{bmatrix} = \mathbf{J}^{-1} \begin{bmatrix} f_\xi^1 & f_\xi^2 & f_\xi^3 \\ f_\eta^1 & f_\eta^2 & f_\eta^3 \end{bmatrix} \tag{112}$$

Let us proceed to second derivatives of a shape function. One can write

$$\begin{aligned}\frac{\partial^2 f}{\partial \xi^2} &= \frac{\partial}{\partial \xi}\left(\frac{\partial f}{\partial \xi}\right) = \frac{\partial}{\partial \xi}\left(\frac{\partial f}{\partial x}\frac{\partial x}{\partial \xi} + \frac{\partial f}{\partial y}\frac{\partial y}{\partial \xi}\right) \\ &= f_x x_{\xi\xi} + x_\xi(f_{xx}x_\xi + f_{xy}y_\xi) + f_y y_{\xi\xi} + y_\xi(f_{xy}x_\xi + f_{yy}y_\xi)\end{aligned} \tag{113}$$

and in the same manner,

$$\frac{\partial^2 f}{\partial \eta^2} = f_x x_{\eta\eta} + x_\eta(f_{xx}x_\eta + f_{xy}y_\eta) + f_y y_{\eta\eta} + y_\eta(f_{xy}x_\eta + f_{yy}y_\eta) \tag{114}$$

$$\frac{\partial^2 f}{\partial \xi \eta} = f_x x_{\xi\eta} + f_y y_{\xi\eta} + x_\xi x_\eta f_{xx} + (x_\xi y_\eta + x_\eta y_\xi) f_{xy} + y_\xi y_\eta f_{yy} \tag{115}$$

Putting equations (113,114,115) in matrix form, one can write

$$\begin{bmatrix} f_{\xi\xi} \\ f_{\eta\eta} \\ f_{\xi\eta} \end{bmatrix} = \begin{bmatrix} x_\xi^2 & y_\xi^2 & 2x_\xi y_\xi \\ x_\eta^2 & y_\eta^2 & 2x_\eta y_\eta \\ x_\xi x_\eta & y_\xi y_\eta & x_\xi y_\eta + x_\eta y_\xi \end{bmatrix} \begin{bmatrix} f_{xx} \\ f_{yy} \\ f_{xy} \end{bmatrix} + \begin{bmatrix} x_{\xi\xi} & y_{\xi\xi} \\ x_{\eta\eta} & y_{\eta\eta} \\ x_{\xi\eta} & y_{\xi\eta} \end{bmatrix} \begin{bmatrix} f_x \\ f_y \end{bmatrix} \tag{116}$$

which allows to compute the second derivatives of f with respect to the physical coordinates as

$$\begin{bmatrix} f_{xx} \\ f_{yy} \\ f_{xy} \end{bmatrix} = \mathbf{J}_{33}^{-1} \left(\begin{bmatrix} f_{\xi\xi} \\ f_{\eta\eta} \\ f_{\xi\eta} \end{bmatrix} - \mathbf{J}_{32} \begin{bmatrix} f_x \\ f_y \end{bmatrix} \right) \tag{117}$$

Boundary conditions In the same manner we have done for the Bernoulli beams, in order to fix the rotation BCs, we simple fix two row of control points at the boundary, see Figure 55.

Symmetry boundary conditions Figure 56 illustrates the use of symmetry boundary conditions when only 1/4 plate is modelled. Along the symmetry lines, the rotation should be zero which can be enforced by constraining the deflection (w) of two rows of control points along these lines together. These contraints can be implemented using a simple penalty technique as shown in Listing 7.1.

Listing 7.1: Enforcing symmetry BCs.

```
w         = 1e7;
penaltyStiffness = w*[1 -1;-1 1];
```

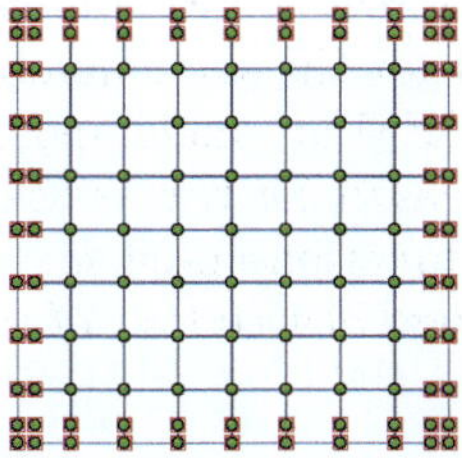

Figure 55: Clamped boundary conditions in a rotation-free IGA formulation: simply fixing the deflections of two rows of control points around the clamped boundary.

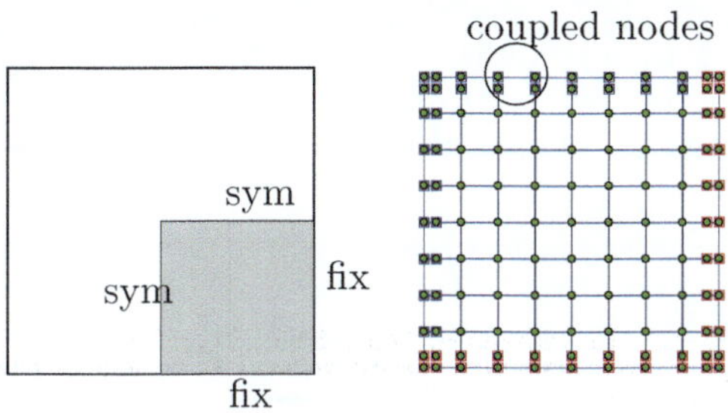

Figure 56: A fully clamped square plate: 1/4 model is analyzied using appropriate symmetry BCs. Along the symmetry lines, the rotation is fixed which can be achieved by enforcing the deflection of two rows of control points that define the tangent of the plate to have the same value.

```
for i=1:length(topNodes)
    sctr   = [topNodes(i) nextToTopNodes(i)];

    K(sctr,sctr) = K(sctr,sctr) + penaltyStiffness;
end
% the same for the left two rows
```

For completeness, we give the implementation of Kirchhoff plate elements in Listing 7.2. Note that we have skipped codes that are common with IGA code for 2D continua.

Listing 7.2: Computation of stiffness matrix for the Kirchhoff plate.

```
for gp=1:size(W,1)
  pt      = Q(gp,:);
  wt      = W(gp);
  %shape functions, 1st/2nd ders. w.r.t natural coords
  [R dRdxi dRdeta dR2dxi dR2det dR2dxe] = ...
       NURBS2DBasis2ndDers([Xi; Eta],p,q,uKnot,vKnot,weights');
  % the derivative w.r.t spatial physical coordinates
  jacob  = [dRdxi; dRdeta]              * pts; % 2x2 matrix
  jacob2 = [dR2dxi; dR2det; dR2dxe] * pts; % 3x2 matrix

  J1     = det(jacob);
  dxdxi  = jacob(1,1);  dydxi  = jacob(1,2);
  dxdet  = jacob(2,1);  dydet  = jacob(2,2);
  j33    = [dxdxi^2      dydxi^2      2*dxdxi*dydxi;
            dxdet^2      dydet^2      2*dxdet*dydet;
            dxdxi*dxdet  dydxi*dydet  dxdxi*dydet+dxdet*dydxi];
  % Jacobian inverse and spatial 1st and 2nd derivatives
  invJacob    = inv(jacob);
  dRdx        = invJacob*[dRdxi;dRdeta];
  dR2dx       = inv(j33)*([dR2dxi; dR2det; dR2dxe]-jacob2*dRdx);
  % B matrix
  B           = dR2dx;
  B(3,:)      = B(3,:)*2;
  % compute elementary stiffness matrix and
  K(sctr,sctr) = K(sctr,sctr) + B' * C * B * J1 * J2 * wt;
end
```

Numerical examples A circular plate subjected to uniformly distributed transverse load is studied. Material properties are $E = 10^7$ and $\nu = 0.3$. The geometry of a circular surface can be described by one bi-cubic NURBS surface with nine control points as shown in Figure 57. A typical mesh is shown in 58. Note that this parametrization is not unique but others are not suitable for numerical analysis albeit geometrically suitable. For a thoughout discussion on geometry parametrization for both design and analysis we refer to Cohen et al. (2010). Figure 59 plots some mode shapes. Readers are referred to files **igaCircularPlate.m** and **igaCircularPlateModalAnalysis.m** for the implementation.

From CAD to FEA: continuity issues We provide a simple example to illustrate the importance of satisfaction of the requirement of C^1 continuity when using the Kirchhoff theory. To this end, consider a cantilever plate that subjects to a vertical force at the right extreme. The plate geometry is described using a bi-cubic B-splines surface in which there are points with

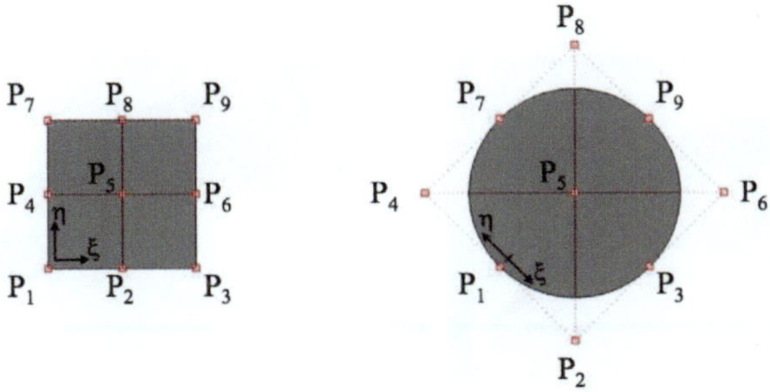

Figure 57: Circular plate modeled by degenerating a square after Kiendle's PhD thesis Kiendl (2011).

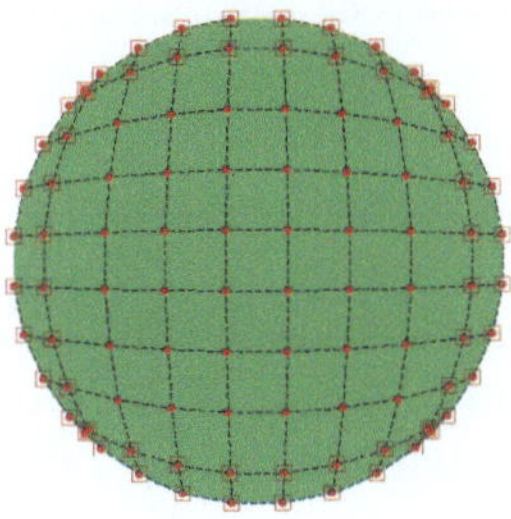

Figure 58: Circular plate: mesh of 8×8 quadratic elements. Also shown are the fixed control points. Two rows to impose the zero rotation BC.

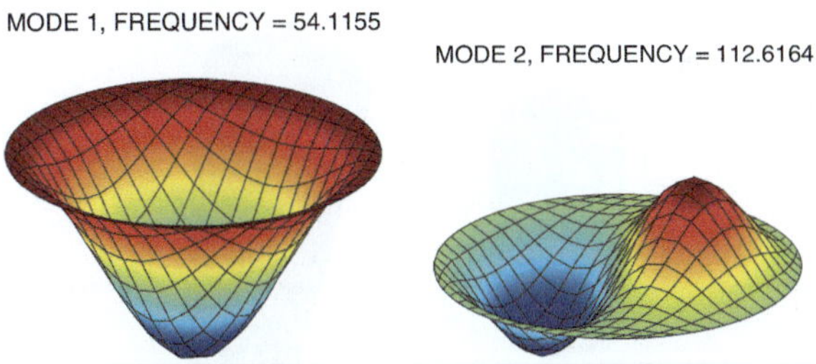

(a) NURBSmesh (b) NURBSmesh

Figure 59: Mode shapes of a clamped circular plate.

only C^0 continuity. We solve this problem by using both Kirchhoff plate element and Mindlin plate element. The result is given in Figure 60.

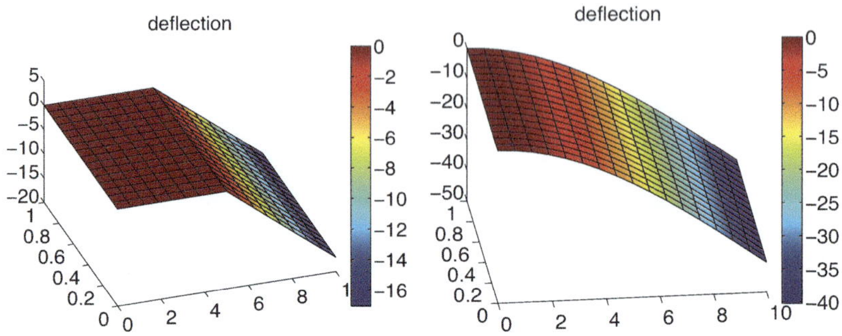

Figure 60: Cantilever beam subjected to a vertical force: in the geometry model there is a row of C^0 control points that result in incorrect behavior modelled by the Kirchhoff theory (left). Note that using the Reissner-Mindlin theory results in a correct response (right).

We emphasize that the existence of C^0 points in NURBS geometries is inavoidable, see e.g., Figure 61. In order to use such geometries, one has to detect C^0 control points and in the analysis, create contraints on them such that they satisfy the geometric continuity G^1. Generally, the parametric continuity C^k implies the geometric continuity G^k but not vice versa. For a discussion on parametric continuity and geometric continuity we refer to Kiendl (2011) and Rogers (2001).

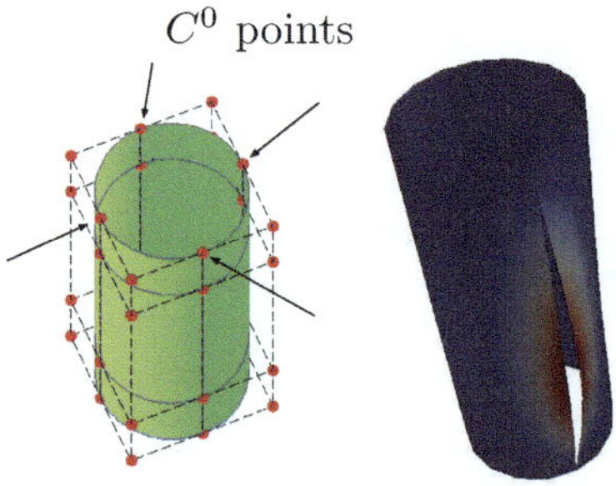

Figure 61: Inavoidable C^0 points in common NURBS objects. Furthermore, for rotational surfaces, there are C^{-1} points that result in a model which is not suitable for analysis.

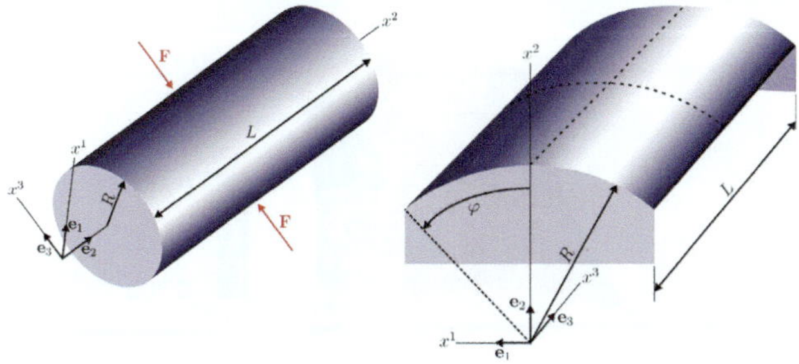

Figure 62: Some benchmark shell problems: pinched cylinder (left) and Scordelis-Lo roof (right) .

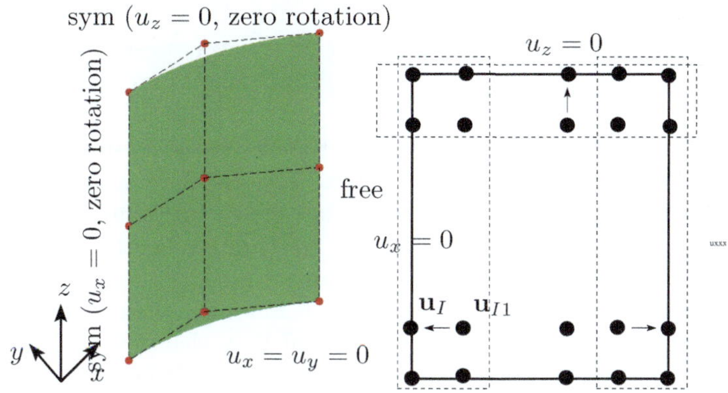

Figure 63: Scordelis Lo roof: 1/4 model with one bi-qudratic NURBS surface. The control points and weights are given in Listing **??**. The z-axis points in the length direction. For symmetry edges, in order to satisfy the symmetry condition i.e., zero rotation, we constraint two rows of control points in the sense that the displacements of the control points right next to the control points locating on the symmetry edges ($\mathbf{u}_{I1}$) to match $\mathbf{u}_I$.

7.4 Numerical examples

In this section some examples on linear and geometrically nonlinear shells are presented to demonstrate the performance of IGA. In order to help those who are not yet familiar with NURBS geometry data are provided.

Scordelis-Lo roof The configuration of this problem is given in Figure 62 (right). Due to symmetry, only 1/4 of the roof is modeled although there

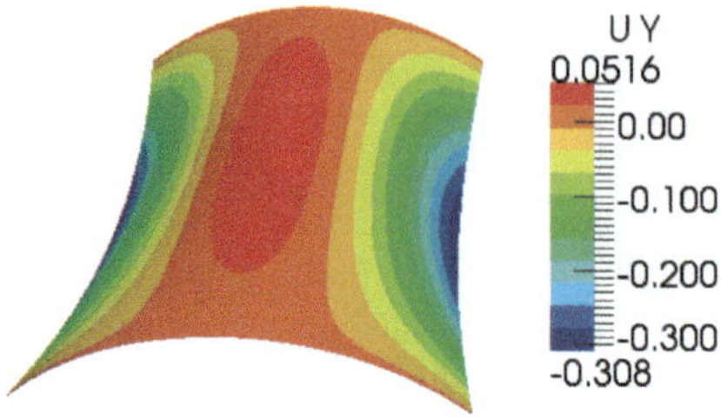

Figure 64: Scordelis Lo roof: contour plot of the y displacement field. The reference solution is -0.3024. And our result with 16×16 bi-cubic elements is -0.3076.

exists a more important reason (rather than saving computational cost) for doing so that will be made clear later. The surface can be described by one single bi-quadric NURBS element as shown in Figure 63. Symmetry conditions are handled using a penalty method, see Section 7.3. Figure 64 shows the contour plot of the y displacement field.

Suppose that one needs to model the full shell then there will be C^0 points which in the framework of Kirchhoff-Love shell formulation requires the G^1 enforcement. If the FEA analyst is not the CAD geometry modeler, then the detection of C^0 points is not a straightforward task.

Pinched cylinder is given in Figure 62. Again, due to symmetry, only 1/4 model is considered, see Figure 65. Refined models are realized by the $k-$ refinement. We refer to Figure 66 for the contour plot of the y displacement on the deformed configuration.

Cantilever shell subjected to an end shear force The problem description is given in Figure 67. Material properties are $E = 1.2 \times 10^6$, $\nu = 0.0$. $t = 0.1$, $L = 10.0$, $b = 1.0$. The shear force remains in the z-direction during the deformation. Matlab implementation is given in file **igaGNLSquareShell.m**. The mesh is given in Figure 68 and the deformed configurations are plotted in Figure 69. The load-displacement is given in Figure 70 with a good agreement with reference result given in Sze et al. (2004).

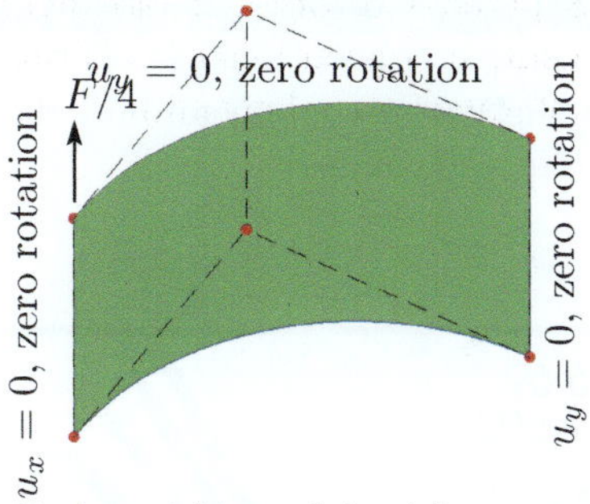

Figure 65: Pinched cylinder: 1/4 model with one qudratic-linear NURBS surface.

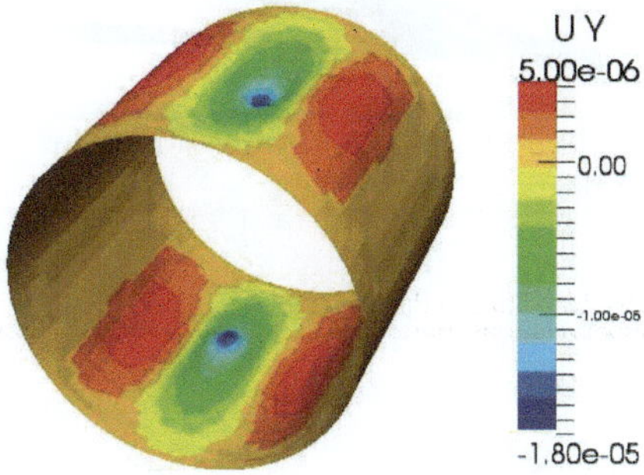

Figure 66: Pinched cylinder: contour plot of the y displacement field.

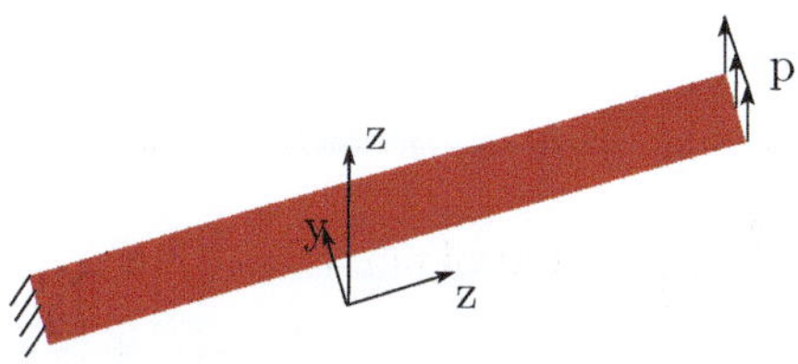

Figure 67: Cantilever shell subjected to an end shear force: setup.

Pull-out of an open-ended cylindrical shell We consider the pull out of an open-ended cylindrical shell as one of the most common benchmarks for geometrically nonlinear thin shell problems Millán et al. (2013); Sze et al.

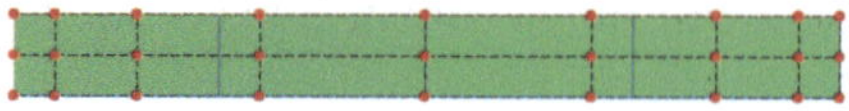

Figure 68: Cantilever shell subjected to an end shear force: a mesh of 4 quintic-quadratic B-splines elements.

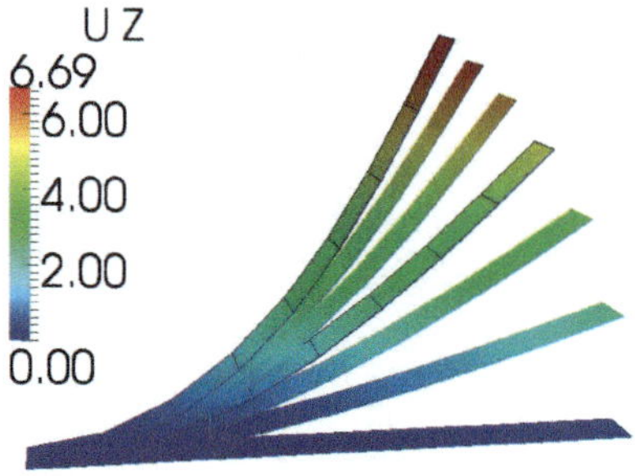

Figure 69: Cantilever shell subjected to an end shear force: deformed configurations without scaling.

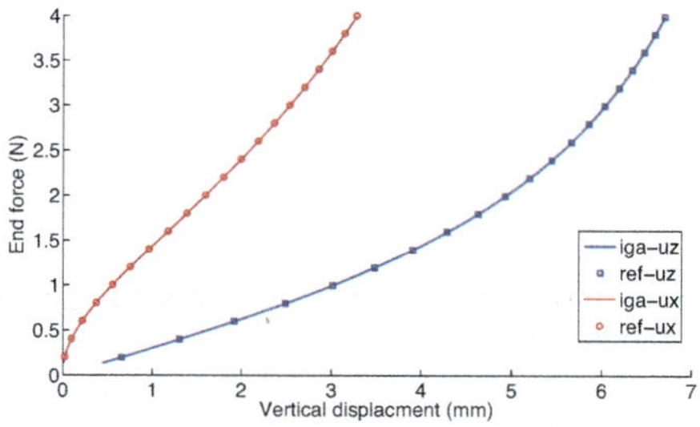

Figure 70: Cantilever shell subjected to an end shear force: load-displacement responses.

(2004). The problem is described in Figure 71. The deformed shape is given in Figure 72 and the load-displacement curve is given in Figure 73.

Acknowledgements

The authors would like to acknowledge the financial support of the Framework Programme 7 Initial Training Network Funding under grant number

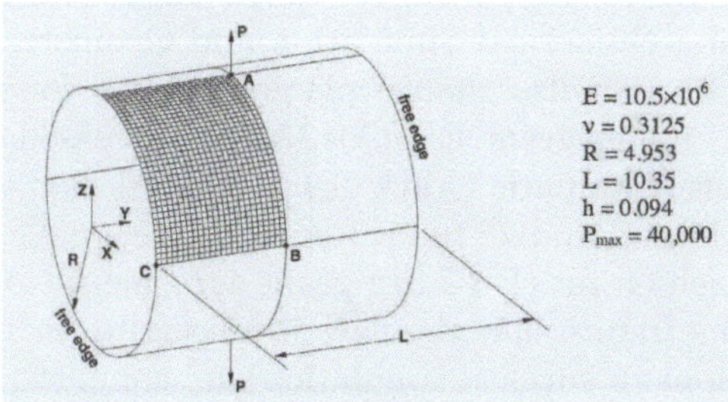

Figure 71: The open-end cylindrical shell subjected to radial pulling forces: problem setup.

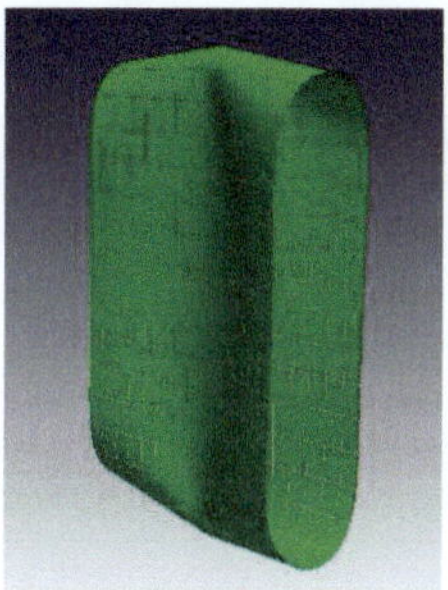

Figure 72: The open-end cylindrical shell subjected to radial pulling forces: deformed shape without scaling.

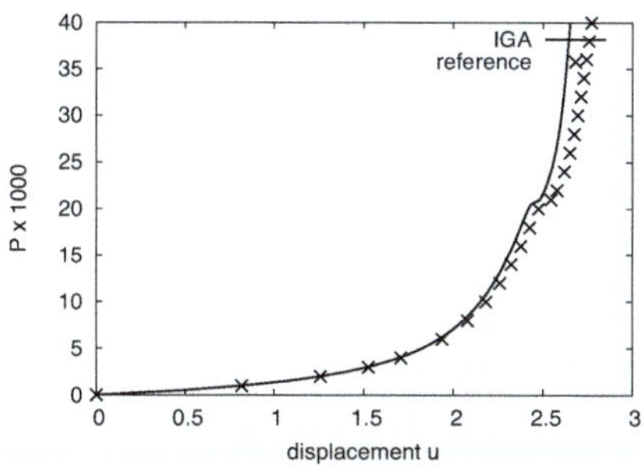

Figure 73: The open-end cylindrical shell subjected to radial pulling forces: load-displacement responses. Reference solution is taken from Sze et al. (2004).

289361 "Integrating Numerical Simulation and Geometric Design Technology". Stéphane Bordas also thanks partial funding for his time provided by 1) the EPSRC under grant EP/G042705/1 Increased Reliability for Industrially Relevant Automatic Crack Growth Simulation with the eXtended Finite Element Method and 2) the European Research Council Starting Independent Research Grant (ERC Stg grant agreement No. 279578) entitled "Towards real time multiscale simulation of cutting in non-linear materials with applications to surgical simulation and computer iMAM at Cardiff University.

Bibliography

M. Aigner, C. Heinrich, B. Jüttler, E. Pilgerstorfer, B. Simeon, and A.-V. Vuong. Swept volume parameterization for isogeometric analysis. In E.R. Hancock, R.R. Martin, and M.A. Sabin, editors, *Mathematics of Surfaces XIII*, volume 5654 of *Lecture Notes in Computer Science*, pages 19–44. Springer Berlin Heidelberg, 2009. ISBN 978-3-642-03595-1.

G. Alfano and M. A. Crisfield. Finite element interface models for the delamination analysis of laminated composites: mechanical and computational is sues. *International Journal for Numerical Methods in Engineering*, 50 (7):1701–1736, 2001.

O. Allix, P. Ladevèze, and A. Corigliano. Damage analysis of interlaminar fracture specimens. *Composite Structures*, 31(1):61 – 74, 1995.

F. Armero and C. Linder. Numerical simulation of dynamic fracture using finite elements with embedded discontinuities. *International Journal of Fracture*, 160(2):119–141, 2009.

F. Auricchio, L. da Veiga Beirão, A. Buffa, C. Lovadina, A. Reali, and G. Sangalli. A fully âĂIJlocking-freeâĂİ isogeometric approach for plane linear elasticity problems: A stream function formulation. *Computer Methods in Applied Mechanics and Engineering*, 197(1-4):160–172, 2007.

F. Auricchio, L. Da Veiga, T.J.R. Hughes, A. Reali, and G. Sangalli. Isogeometric collocation methods. *Mathematical Models and Methods in Applied Sciences*, 20(11):2075–2107, 2010.

F. Auricchio, F. Calabro, T.J.R. Hughes, A. Reali, and G. Sangalli. A simple algorithm for obtaining nearly optimal quadrature rules for NURBS-based isogeometric analysis. *Computer Methods in Applied Mechanics and Engineering*, 249–252:15 – 27, 2012a.

F. Auricchio, L. BeirÃČÂčo da Veiga, T.J.R. Hughes, A. Reali, and G. Sangalli. Isogeometric collocation for elastostatics and explicit dynamics. *Computer Methods in Applied Mechanics and Engineering*, 249–252:2 – 14, 2012b.

Y. Bazilevs and I. Akkerman. Large eddy simulation of turbulent Taylor-Couette flow using isogeometric analysis and the residual-based variational multiscale method. *Journal of Computational Physics*, 229(9): 3402–3414, 2010.

Y. Bazilevs, V. M. Calo, T. J. R. Hughes, and Y. Zhang. Isogeometric fluid-structure interaction: theory, algorithms, and computations. *Computational Mechanics*, 43:3–37, 2008.

Y. Bazilevs, J.R. Gohean, T.J.R. Hughes, R.D. Moser, and Y. Zhang. Patient-specific isogeometric fluid-structure interaction analysis of thoracic aortic blood flow due to implantation of the Jarvik 2000 left ventricular assist device. *Computer Methods in Applied Mechanics and Engineering*, 198(45-46):3534–3550, 2009.

Y. Bazilevs, V.M. Calo, J.A. Cottrell, J.A. Evans, T.J.R. Hughes, S. Lipton, M.A. Scott, and T.W. Sederberg. Isogeometric analysis using T-splines. *Computer Methods in Applied Mechanics and Engineering*, 199(5-8):229–263, 2010.

L. Beirão da Veiga, A. Buffa, C. Lovadina, M. Martinelli, and G. Sangalli. An isogeometric method for the Reissner-Mindlin plate bending problem. *Computer Methods in Applied Mechanics and Engineering*, 209–212:45–53, 2012.

T. Belytschko, Y. Y. Lu, and L. Gu. Element-free galerkin methods. *International Journal for Numerical Methods in Engineering*, 37(2):229–256, 1994a.

T. Belytschko, Y. Y. Lu, and L. Gu. Element-free Galerkin methods. *International Journal for Numerical Methods in Engineering*, 37:229–256, 1994b.

D. J. Benson, Y. Bazilevs, E. De Luycker, M.-C. Hsu, M. Scott, T. J. R. Hughes, and T. Belytschko. A generalized finite element formulation for arbitrary basis functions: From isogeometric analysis to XFEM. *International Journal for Numerical Methods in Engineering*, 83(6):765–785, 2010a.

D.J. Benson, Y. Bazilevs, M.C. Hsu, and T.J.R. Hughes. Isogeometric shell analysis: The Reissner–Mindlin shell. *Computer Methods in Applied Mechanics and Engineering*, 199(5-8):276–289, 2010b.

D.J. Benson, Y. Bazilevs, M.-C. Hsu, and T.J.R. Hughes. A large deformation, rotation-free, isogeometric shell. *Computer Methods in Applied Mechanics and Engineering*, 200(13-16):1367–1378, 2011.

D.J. Benson, S. Hartmann, Y. Bazilevs, M.-C. Hsu, and T.J.R. Hughes. Blended isogeometric shells. *Computer Methods in Applied Mechanics and Engineering*, 255:133 – 146, 2013.

S. Bordas, P. V. Nguyen, C. Dunant, A. Guidoum, and H. Nguyen-Dang. An extended finite element library. *International Journal for Numerical Methods in Engineering*, 71(6):703–732, 2007.

M. J. Borden, M. A. Scott, J. A. Evans, and T. J. R. Hughes. Isogeometric finite element data structures based on Bézier extraction of NURBS. *International Journal for Numerical Methods in Engineering*, 87(1âĂŘ5): 15–47, 2011.

M. J. Borden, C. V. Verhoosel, M. A. Scott, T. J.R. Hughes, and C. M. Landis. A phase-field description of dynamic brittle fracture. *Computer Methods in Applied Mechanics and Engineering*, 217–220:77 – 95, 2012.

P.B. Bornemann and F. Cirak. A subdivision-based implementation of the hierarchical b-spline finite element method. *Computer Methods in Applied Mechanics and Engineering*, 253:584 – 598, 2013.

A. Buffa, D. Cho, and G. Sangalli. Linear independence of the T-spline blending functions associated with some particular T-meshes. *Computer Methods in Applied Mechanics and Engineering*, 1437-1445:199, 2010.

D. Burkhart, B. Hamann, and G. Umlauf. Iso-geometric Finite Element Analysis Based on Catmull-Clark : subdivision Solids. *Computer Graphics Forum*, 29(5):1575–1584, 2010.

G.T. Camacho and M. Ortiz. Computational modelling of impact damage in brittle materials. *International Journal of Solids and Structures*, 33 (20âĂŞ22):2899 – 2938, 1996.

Rui P. R. Cardoso and J. M. A. Cesar de Sa. The enhanced assumed strain method for the isogeometric analysis of nearly incompressible deformation of solids. *International Journal for Numerical Methods in Engineering*, 92(1):56–78, 2012.

J. Chessa. Programming the finite element method with Matlab. Northwestern University, http://www.tam.northwestern.edu/jfc795/Matlab/, 2002. URL `http://www.tam.northwestern.edu/jfc795/Matlab/`.

E. Cohen, T. Martin, R.M. Kirby, T. Lyche, and R.F. Riesenfeld. Analysis-aware modeling: Understanding quality considerations in modeling for isogeometric analysis. *Computer Methods in Applied Mechanics and Engineering*, 199(5-8):334–356, 2010.

N. Collier, D. Pardo, L. Dalcin, M. Paszynski, and V.M. Calo. The cost of continuity: A study of the performance of isogeometric finite elements using direct solvers. *Computer Methods in Applied Mechanics and Engineering*, 213–216:353 – 361, 2012.

J. A. Cottrell, T. J.R. Hughes, and Y. Bazilevs. *Isogeometric Analysis: Toward Integration of CAD and FEA*. Wiley, 2009. ISBN 9780470748372.

D. Dias da Costa, J. Alfaiate, L.J. Sluys, and E. Júlio. A discrete strong discontinuity approach. *Engineering Fracture Mechanics*, 76(9):1176 – 1201, 2009.

C. de Falco, A. Reali, and R. Vázquez. GeoPDEs: a research tool for Isogeometric Analysis of PDEs. *Advances in Engineering Software*, 42 (12):1020–1034, 2011.

L. De Lorenzis, İ. Temizer, P. Wriggers, and G. Zavarise. A large deformation frictional contact formulation using NURBS-bases isogeometric analysis. *International Journal for Numerical Methods in Engineering*, 87(13):1278–1300, 2011.

E. De Luycker, D. J. Benson, T. Belytschko, Y. Bazilevs, and M. C. Hsu. X-FEM in isogeometric analysis for linear fracture mechanics. *International Journal for Numerical Methods in Engineering*, 87(6):541–565, 2011.

J. Deng, F. Chen, X. Li, C. Hu, W. Tong, Z. Yang, and Y. Feng. Polynomial splines over hierarchical T-meshes. *Graphical Models*, 70:76–86, 2008.

D. Dias-da Costa, J. Alfaiate, L.J. Sluys, and E. Júlio. A comparative study on the modelling of discontinuous fracture by means of enriched nodal and element techniques and interface elements. *International Journal of Fracture*, 161(1):97–119, 2010. ISSN 0376-9429.

T. Dokken and V. Skytt. Locally refined splines. In *Proceedings of IV European Conference On Computational Mechanics. Solids, Structures and Coupled Problems in Engineering*, Paris, France, 16-21 May 2010.

M. R. Dörfel, B. Jüttler, and B. Simeon. Adaptive isogeometric analysis by local h-refinement with T-splines. *Computer Methods in Applied Mechanics and Engineering*, 199(5-8):264–275, 2010.

R. Echter and M. Bischoff. Numerical efficiency, locking and unlocking of NURBS finite elements. *Computer Methods in Applied Mechanics and Engineering*, 199(5–8):374 – 382, 2010.

R. Echter, B. Oesterle, and M. Bischoff. A hierarchic family of isogeometric shell finite elements. *Computer Methods in Applied Mechanics and Engineering*, 254:170 – 180, 2013.

T. Elguedj, Y. Bazilevs, V.M. Calo, and T.J.R. Hughes. B-bar and f-bar projection methods for nearly incompressible linear and non-linear elasticity and plasticity using higher-order NURBS elements. *Computer Methods in Applied Mechanics and Engineering*, 197(33-40):2732–2762, 2008a.

T. Elguedj, Y. Bazilevs, V.M. Calo, and T.J.R. Hughes. F-bar projection method for finite deformation elasticity and plasticity using nurbs based isogeometric analysis. *International Journal of Material Forming*, 1(1): 1091–1094, 2008b.

A. Embar, J. Dolbow, and I. Harari. Imposing Dirichlet boundary conditions with Nitsche's method and spline-based finite elements. *International Journal for Numerical Methods in Engineering*, 83(7):877–898, 2010.

J.M. Escobar, J.M. Cascón, E. Rodriguez, and R. Montenegro. A new approach to solid modeling with trivariate T-splines based on mesh optimization. *Computer Methods in Applied Mechanics and Engineering*, 200(45-46):3210–3222, 2011.

H. D. Espinosa, P. D. Zavattieri, and S. K. Dwivedi. A finite deformation continuum-discrete model for the description of fragmentation and damage in brittle materials. *Journal of the Mechanics and Physics of Solids*, 46(10):1909 – 1942, 1998.

P. Fischer, M. Klassen, J. Mergheim, P. Steinmann, and R. Müller. Isogeometric analysis of 2D gradient elasticity. *Computational Mechanics*, 47: 325–334, 2010.

T. P. Fries and T. Belytschko. The extended/generalized finite element method: An overview of the method and its applications. *International Journal for Numerical Methods in Engineering*, 84(3):253–304, 2010.

S. S. Ghorashi, N. Valizadeh, and S. Mohammadi. Extended isogeometric analysis for simulation of stationary and propagating cracks. *International Journal for Numerical Methods in Engineering*, 2012. In Press.

H. Gómez, V. M. Calo, Y. Bazilevs, and T.J.R. Hughes. Isogeometric analysis of the Cahn-Hilliard phase-field model. *Computer Methods in Applied Mechanics and Engineering*, 197(49-50):4333–4352, 2008.

H. Gomez, T.J.R. Hughes, X. Nogueira, and V. M. Calo. Isogeometric analysis of the isothermal Navier-Stokes-Korteweg equations. *Computer Methods in Applied Mechanics and Engineering*, 199(25-28):1828–1840, 2010.

A. Henderson. *ParaView Guide, A Parallel Visualization Application*. Kitware Inc., 2007.

T.J.R. Hughes, J.A. Cottrell, and Y. Bazilevs. Isogeometric analysis: CAD, finite elements, NURBS, exact geometry and mesh refinement. *Computer Methods in Applied Mechanics and Engineering*, 194(39-41):4135–4195, 2005.

T.J.R. Hughes, A. Reali, and G. Sangalli. Efficient quadrature for NURBS-based isogeometric analysis. *Computer Methods in Applied Mechanics and Engineering*, 199(5-8):301–313, 2010.

Lu J. Circular element: Isogeometric elements of smooth boundary. *Computer Methods in Applied Mechanics and Engineering*, 198(30-32):2391–2402, 2009.

L. Jia. Isogeometric contact analysis: Geometric basis and formulation for frictionless contact. *Computer Methods in Applied Mechanics and Engineering*, 200(5-8):726–741, 2011.

P. Kerfriden, O. Allix, and P. Gosselet. A three-scale domain decomposition method for the 3D analysis of debonding in laminates. *Computational Mechanics*, 44(3):343 – 362, 2009.

J. Kiendl, K.-U. Bletzinger, J. Linhard, and R. Wüchner. Isogeometric shell analysis with Kirchhoff-Love elements. *Computer Methods in Applied Mechanics and Engineering*, 198(49-52):3902–3914, 2009.

J. Kiendl, Y. Bazilevs, M.-C. Hsu, R. Wüchner, and K.-U. Bletzinger. The bending strip method for isogeometric analysis of Kirchhoff-Love shell structures comprised of multiple patches. *Computer Methods in Applied Mechanics and Engineering*, 199(37-40):2403–2416, 2010.

J. M. Kiendl. *Isogeometric Analysis and Shape Optimal Design of Shell Structures*. PhD thesis, Technische universität münchen, 2011.

Hyun-Jung Kim, Yu-Deok Seo, and Sung-Kie Youn. Isogeometric analysis with trimming technique for problems of arbitrary complex topology. *Computer Methods in Applied Mechanics and Engineering*, 199(45-48): 2796–2812, 2010.

S. K. Kleiss, B. Jüttler, and W. Zulehner. Enhancing isogeometric analysis by a finite element-based local refinement strategy. *Computer Methods in Applied Mechanics and Engineering*, 213–216:168 – 182, 2012.

S. Lipton, J.A. Evans, Y. Bazilevs, T. Elguedj, and T.J.R. Hughes. Robustness of isogeometric structural discretizations under severe mesh distortion. *Computer Methods in Applied Mechanics and Engineering*, 199 (5–8):357 – 373, 2010.

J. Lu and X. Zhou. Cylindrical element: Isogeometric model of continuum rod. *Computer Methods in Applied Mechanics and Engineering*, 200(1-4): 233–241, 2011.

N. D. Manh, A. Evgrafov, A. R. Gersborg, and J. Gravesen. Isogeometric shape optimization of vibrating membranes. *Computer Methods in Applied Mechanics and Engineering*, 200(13-16):1343–1353, 2011.

T. Martin, E. Cohen, and R. M. Kirby. Volumetric parameterization and trivariate B-spline fitting using harmonic functions. *Computer Aided Geometric Design*, 26(6):648–664, 2009.

A. Masud and R. Kannan. B-splines and NURBS based finite element methods for Kohn-Sham equations. *Computer Methods in Applied Mechanics and Engineering*, 241-244:112 – 127, 2012.

M.E. Matzen, T. Cichosz, and M. Bischoff. A point to segment contact formulation for isogeometric, NURBS based finite elements. *Computer Methods in Applied Mechanics and Engineering*, 255:27 – 39, 2013.

D. Millán, A. Rosolen, and M. Arroyo. Nonlinear manifold learning for meshfree finite deformation thin-shell analysis. *International Journal for Numerical Methods in Engineering*, 93(7):685–713, 2013.

N. Moës, J. Dolbow, and T. Belytschko. A finite element method for crack growth without remeshing. *International Journal for Numerical Methods in Engineering*, 46(1):131–150, 1999.

B. Nayroles, G. Touzot, and P. Villon. Generalizing the finite element method: Diffuse approximation and diffuse elements. *Computational Mechanics*, 10(5):307–318, 1992.

V. P. Nguyen and H. Nguyen-Xuan. High-order B-splines based finite elements for delamination analysis of laminated composites. *Composite Structures*, 102:261–275, 2013.

V. P. Nguyen, T. Rabczuk, S. Bordas, and M. Duflot. Meshless methods: A review and computer implementation aspects. *Mathematics and Computers in Simulation*, 79(3):763–813, 2008.

V. P. Nguyen, P. Kerfriden, and S. Bordas. Two- and three-dimensional isogeometric cohesive elements for composite delamination analysis. *Composites Part B*, 60:193–212, 2014a.

V.P. Nguyen. Discontinuous Galerkin/Extrinsic cohesive zone modeling: implementation caveats and applications in computational fracture mechanics. *Engineering Fracture Mechanics*, 128:37–68, 2014.

V.P. Nguyen, P. Kerfriden, S.P.A. Bordas, and T. Rabczuk. Isogeometric analysis suitable trivariate NURBS representation of composite panels with a new offset algorit hm. *Computer Aided Design*, 55:49–63, 2014b.

V.P. Nguyen, P. Kerfriden, M. Brino, S. Bordas, and E. Bonisoli. Nitsche's method for two and three dimensional NURBS patch coupling. *Computational Mechanics*, 53(6):1163–1182, 2014c.

N. Nguyen-Thanh, J. Kiendl, H. Nguyen-Xuan, R. Wüchner, K.U. Bletzinger, Y. Bazilevs, and T. Rabczuk. Rotation free isogeometric thin shell analysis using PHT-splines. *Computer Methods in Applied Mechanics and Engineering*, 200(47-48):3410–3424, 2011a.

N. Nguyen-Thanh, H. Nguyen-Xuan, S.P.A. Bordas, and T. Rabczuk. Isogeometric analysis using polynomial splines over hierarchical T-meshes for two-dimensional elastic solids. *Computer Methods in Applied Mechanics and Engineering*, 200(21-22):1892–1908, 2011b.

P. N. Nielsen, A. R. Gersborg, J. Gravesen, and N. L. Pedersen. Discretizations in isogeometric analysis of Navier-Stokes flow. *Computer Methods in Applied Mechanics and Engineering*, 200(45-46):3242–3253, 2011.

A. Pandolfi and M. Ortiz. An efficient adaptive procedure for three-dimensional fragmentation simulations. *Engineering with Computers*, 18(2):148–159, 2002.

K. Park, G. H. Paulino, W. Celes, and R. Espinha. Adaptive mesh refinement and coarsening for cohesive zone modeling of dynamic fracture. *International Journal for Numerical Methods in Engineering*, 92(1):1–35, 2012.

M.J. Peake, J. Trevelyan, and G. Coates. Extended isogeometric boundary element method (xibem) for two-dimensional helmholtz problems. *Computer Methods in Applied Mechanics and Engineering*, 259:93 – 102, 2013.

L. A. Piegl and W. Tiller. *The NURBS Book*. Springer, 1996. ISBN 3540615458.

X. Qian. Full analytical sensitivities in NURBS based isogeometric shape optimization. *Computer Methods in Applied Mechanics and Engineering*, 199(29-32):2059–2071, 2010.

X. Qian and O. Sigmund. Isogeometric shape optimization of photonic crystals via Coons patches. *Computer Methods in Applied Mechanics and Engineering*, 200(25-28):2237–2255, 2011.

T. Rabczuk. Computational methods for fracture in brittle and quasi-brittle solids: State-of-the-art review and future perspectives. *ISRN Applied Mathematics*, 2013. doi: 10.1155/2013/849231. Article ID 849231, 38 pages.

T. Rabczuk, R. Gracie, J. H. Song, and T. Belytschko. Immersed particle method for fluid-structure interaction. *International Journal for Numerical Methods in Engineering*, 81(1):48–71, 2010.

J. N. Reddy. *An Introduction to the Finite Element Method*. McGraw-Hill Education, 3rd edition, 2005.

D. F. Rogers. *An Introduction to NURBS with Historical Perspective*. Academic Press, 2001.

M. Ruess, D. Schillinger, Y. Bazilevs, V. Varduhn, and E. Rank. Weakly enforced essential boundary conditions for NURBS-embedded and trimmed NURBS geometries on the basis of the finite cell method. *International Journal for Numerical Methods in Engineering*, pages n/a–n/a, 2013.

D. Rypl and B. Patzák. From the finite element analysis to the isogeometric analysis in an object oriented computing environment. *Advances in Engineering Software*, 44:116–125, 2012.

J.C.J. Schellekens and R. De Borst. A non-linear finite element approach for the analysis of mode-I free edge delamination in composites. *International Journal of Solids and Structures*, 30(9):1239 – 1253, 1993.

D. Schillinger, L. Dedé, M. A. Scott, J. A. Evans, M. J. Borden, E. Rank, and T. J. R. Hughes. An Isogeometric Design-through-analysis Methodology based on Adaptive Hierarchical Refinement of NURBS, Immersed Boundary Methods, and T-spline CAD Surfaces. *Computer Methods in Applied Mechanics and Engineering*, 242-252:116–150, 2012.

R. Schmidt, J. Kiendl, K.-U. Bletzinger, and R. Wüchner. Realization of an integrated structural design process: analysis-suitable geometric modelling and isogeometric analysis. *Computing and Visualization in Science*, 13(7):315–330, 2010.

R. Schmidt, R. Wuchner, and K.U. Bletzinger. Isogeometric analysis of trimmed NURBS geometries. *Computer Methods in Applied Mechanics and Engineering*, 241–244:93 – 111, 2012.

M. A. Scott, M. J. Borden, C. V. Verhoosel, T. W. Sederberg, and T. J. R. Hughes. Isogeometric finite element data structures based on Bézier extraction of T-splines. *International Journal for Numerical Methods in Engineering*, 88(2):126–156, 2011.

M.A. Scott, X. Li, T.W. Sederberg, and T.J.R. Hughes. Local refinement of analysis suitable T-splines. *Computer Methods in Applied Mechanics and Engineering*, 213-216:206–222, 2012.

M.A. Scott, R.N. Simpson, J.A. Evans, S. Lipton, S.P.A. Bordas, T.J.R. Hughes, and T.W. Sederberg. Isogeometric boundary element analysis using unstructured T-splines. *Computer Methods in Applied Mechanics and Engineering*, 254:197 – 221, 2013.

T. W. Sederberg, J. Zheng, A. Bakenov, and A. Nasri. T-splines and T-NURCCs. *ACM Transactions on Graphics*, 22:477–484, 2003.

R. Sevilla, S. Fernández-Méndez, and A. Huerta. NURBS-enhanced finite element method (NEFEM). *International Journal for Numerical Methods in Engineering*, 76(1):56–83, 2008.

R. Sevilla, S. Fernández-Méndez, and A. Huerta. 3D NURBS-enhanced finite element method (NEFEM). *International Journal for Numerical Methods in Engineering*, 88(2):103–125, 2011.

J. Shi, D. Chopp, J. Lua, N. Sukumar, and T. Belytschko. Abaqus implementation of extended finite element method using a level set representation of three-dimensional fatigue crack growth and life predictions. *Engineering Fracture Mechanics*, 77:2840–2863, 2010.

J. C. Simo, J. Oliver, and F. Armero. An analysis of strong discontinuities induced by strain-softening in rate-independent inelastic solids. *Computational Mechanics*, 12(5):277–296, 1993.

R.N. Simpson, S.P.A. Bordas, J. Trevelyan, and T. Rabczuk. A two-dimensional isogeometric boundary element method for elastostatic analysis. *Computer Methods in Applied Mechanics and Engineering*, 209–212: 87–100, 2012.

R.N. Simpson, S.P.A. Bordas, H. Lian, and J. Trevelyan. An isogeometric boundary element method for elastostatic analysis: 2D implementation aspects. *Computers & Structures*, 118:2 – 12, 2013.

J.H. Song, H. Wang, and T. Belytschko. A comparative study on finite element methods for dynamic fracture. *Computational Mechanics*, 42 (2):239–250, 2008.

H. Speleers, C. Manni, F. Pelosi, and M. L. Sampoli. Isogeometric analysis with Powell-Sabin splines for advection-diffusion-reaction problems. *Computer Methods in Applied Mechanics and Engineering*, 221–222:132 – 148, 2012.

T. Strouboulis, K. Copps, and I. BabuÅąka. The generalized finite element method. *Computer Methods in Applied Mechanics and Engineering*, 190 (32âĂŞ33):4081 – 4193, 2001.

N. Sukumar, D. L. Chopp, N. Moës, and T. Belytschko. Modelling holes and inclusions by level sets in the extended finite element method. *Computer Methods in Applied Mechanics and Engineering*, 190:6183–6200, 2000.

K.Y. Sze, X.H. Liu, and S.H. Lo. Popular benchmark problems for geometric nonlinear analysis of shells. *Finite Elements in Analysis and Design*, 40 (11):1551 – 1569, 2004.

H. P. Tada, P. C. Paris, and G. R. Irwin. *The Stress Analysis of Cracks Handbook*. Del Research Corporation, St. Louis, MO, 1985.

T. Takacs and B. Jüttler. Existence of stiffness matrix integrals for singularly parameterized domains in isogeometric analysis. *Computer Methods in Applied Mechanics and Engineering*, 200(49-52):3568–3582, 2011.

A. Tambat and G. Subbarayan. Isogeometric enriched field approximations. *Computer Methods in Applied Mechanics and Engineering*, 245–246:1 – 21, 2012.

R. L. Taylor. Isogeometric analysis of nearly incompressible solids. *International Journal for Numerical Methods in Engineering*, 87(1-5):273–288, 2011.

İ. Temizer, P. Wriggers, and T.J.R. Hughes. Contact treatment in isogeometric analysis with NURBS. *Computer Methods in Applied Mechanics and Engineering*, 200(9-12):1100–1112, 2011.

İ. Temizer, P. Wriggers, and T.J.R. Hughes. Three-Dimensional Mortar-Based frictional contact treatment in isogeometric analysis with NURBS. *Computer Methods in Applied Mechanics and Engineering*, 209–212:115–128, 2012.

T. K. Uhm and S. K. Youn. T-spline finite element method for the analysis of shell structures. *International Journal for Numerical Methods in Engineering*, 80(4):507–536, 2009.

C. V. Verhoosel, M. A. Scott, R. de Borst, and T. J. R. Hughes. An isogeometric approach to cohesive zone modeling. *International Journal for Numerical Methods in Engineering*, 87(1âĂŘ5):336–360, 2011a.

C. V. Verhoosel, M. A. Scott, T. J. R. Hughes, and R. de Borst. An isogeometric analysis approach to gradient damage models. *International Journal for Numerical Methods in Engineering*, 86(1):115–134, 2011b.

A.-V. Vuong, Ch. Heinrich, and B. Simeon. ISOGAT: a 2D tutorial MATLAB code for Isogeometric Analysis. *Computer Aided Geometric Design*, 27(8):644–655, 2010.

A.-V. Vuong, C. Giannelli, B. Jüttler, and B. Simeon. A hierarchical approach to adaptive local refinement in isogeometric analysis. *Computer Methods in Applied Mechanics and Engineering*, 200(49-52):3554–3567, 2011.

W. A. Wall, M. A. Frenzel, and C. Cyron. Isogeometric structural shape optimization. *Computer Methods in Applied Mechanics and Engineering*, 197(33-40):2976–2988, 2008.

D. Wang and J. Xuan. An improved NURBS-based isogeometric analysis with enhanced treatment of essential boundary conditions. *Computer Methods in Applied Mechanics and Engineering*, 199(37-40):2425–2436, 2010.

H. Wang, Y. He, X. Li, X. Gu, and H. Qin. Polycube splines. *Computer-Aided Design*, 40:721–733, 2008.

A. Wawrzinek, K. Hildebrandt, and K. Polthier. Koiter's thin shells on catmull-clark limit surfaces. In *Proceedings of the Vision, Modeling, and Visualization Workshop*, Berlin, Germany, 4-6 October 2011.

G. Xu, B. Mourrain, R. Duvigneau, and A. Galligo. Parameterization of computational domain in isogeometric analysis: Methods and comparison. *Computer Methods in Applied Mechanics and Engineering*, 200 (23-24):2021–2031, 2011.

X.P. Xu and A. Needleman. Numerical simulations of fast crack growth in brittle solids. *Journal of the Mechanics and Physics of Solids*, 42(9), 1994.

Y. Zhang, Y. Bazilevs, S. Goswami, C. L. Bajaj, and T.J.R. Hughes. Patient-specific vascular NURBS modeling for isogeometric analysis of blood flow. *Computer Methods in Applied Mechanics and Engineering*, 196(29-30):2943–2959, 2007a.

Y. Zhang, W. Wang, and T.J.R. Hughes. Solid T-spline construction from boundary representations for genus-zero geometry. *Computer Methods in Applied Mechanics and Engineering*, 249–252:185 – 197, 2012a.

Y. Zhang, W. Wang, and T.J.R. Hughes. Conformal solid T-spline construction from boundary T-spline representations. *Computational Mechanics*, pages 1–9, 2012b.

Z. Zhang, G. H. Paulino, and W. Celes. Extrinsic cohesive modelling of dynamic fracture and microbranching instability in brittle materials. *International Journal for Numerical Methods in Engineering*, 72(8):893–923, 2007b.

Xianlian Zhou and Jia Lu. Nurbs-based galerkin method and application to skeletal muscle modeling. In *Proceedings of the 2005 ACM symposium on Solid and physical modeling*, SPM '05, pages 71–78, New York, NY, USA, 2005. ACM. doi: 10.1145/1060244.1060253.

Boundary Element Methods

Gernot Beer * and Benjamin Marussig †

* Graz University of Technology, Austria and University of Newcastle, Australia
† Graz University of Technology, Austria

Abstract This chapter provides an introduction to the isogeometric Boundary Element Method (BEM). The standard iso-geometric BEM is presented first and then isometric concepts are introduced. Both plane and 3-D problems are discussed and details of implementation given. The method is extended to non-homogeneous and non-linear problems. The main applications are in geotechnical engineering.

1 Introduction

The Boundary Element Method relies on a surface representation of the geometry and is therefore ideally suited to the approach used in computer aided design (CAD) and isogeometric concepts. It means that the description of the problem geometry can be taken directly from CAD data, without the need for mesh generation. The method uses fundamental solutions of the governing differential equations and this means that compatibility and equilibrium conditions are exactly satisfied inside the domain. Therefore, the results are expected to be of higher quality as compared to domain methods. The use of NURBS or B-splines as the basis functions for describing the boundary values results in a further improvement in efficiency and accuracy, as will be shown here.

There has been a misconception that the method is only suitable to analyze linear problems in elasticity and that it is restricted to homogeneous problems. At the end of this chapter it will be shown that with little additional effort such problems can be solved.

G. Beer, S. Bordas (Eds.), *Isogeometric Methods for Numerical Simulation*,
CISM International Centre for Mechanical Sciences
DOI 10.1007/ 978-3-7091-1843-6_3

2 Introduction to the iso-parametric BEM for plane problems

2.1 Boundary integral equations

The boundary integral equations for an elastic continuum without body forces (to be introduced later) can be written as (see Beer et al. (2008)),

$$\mathbf{c}(P)\,\mathbf{u}(P) = \int_S \mathbf{U}(P,Q)\,\mathbf{t}(Q)\,dS - \int_S \mathbf{T}(P,Q)\,\mathbf{u}(Q)\,dS \qquad (1)$$

where P (coordinates x_P, y_P) is the source point and Q (coordinates x_Q, y_Q) is the field point on the boundary S. The coefficient $\mathbf{c}(P)$ is a free term matrix related to the boundary geometry. For smooth boundaries it is given by

$$\mathbf{c} = \frac{1}{2}\mathbf{I} \qquad (2)$$

where $\mathbf{I}$ is the unit matrix. $\mathbf{u}(Q)$ and $\mathbf{t}(Q)$ are vectors containing displacements and tractions on the boundary and $\mathbf{U}(P,Q)$ and $\mathbf{T}(P,Q)$ are matrices containing Kelvin's fundamental solutions (*Kernels*) for the displacements and tractions respectively. For a plane problem we have

$$\mathbf{U}(P,Q) = C\,ln\left(\frac{1}{r}\right)(C_1\mathbf{I} + \mathbf{R}) \qquad (3)$$

where C and C_1 are material constants, r is the distance between point P and Q and

$$\mathbf{R} = \begin{pmatrix} r_x^2 & r_x r_y \\ r_y r_x & r_y^2 \end{pmatrix} \qquad (4)$$

where

$$\begin{aligned} r_x &= \frac{1}{r}(x_Q - x_P) \\ r_y &= \frac{1}{r}(y_Q - y_P) \end{aligned} \qquad (5)$$

As expected $\mathbf{U}$ is symmetric.

The fundamental solution for the tractions is given by

$$\mathbf{T}(P,Q) = C_2\,\frac{1}{r}((C_3\mathbf{I} + 2\,\mathbf{R})\,\cos\theta + C_3\mathbf{R}_1) \qquad (6)$$

where

$$\cos\theta = n_x r_x + n_y r_y \qquad (7)$$

is the angle between the position vector $\mathbf{r}$ and the vector normal to the boundary $\mathbf{n}$ (components n_x, n_y), furthermore

$$\mathbf{R}_1 = \begin{pmatrix} 0 & n_x r_y - n_y r_x \\ n_y r_x - n_x r_y & 0 \end{pmatrix} \tag{8}$$

It can be seen that $\mathbf{T}$ consists of a symmetric part multiplied by $\cos\theta$ and an antisymmetric part represented by $\mathbf{R}_1$.

Both fundamental solutions are singular, i.e. tend to infinity as the source point (P) and the field point (Q) coincide. The first one has a singularity of $o(\ln \frac{1}{r})$ and is therefore *weakly singular*, the second has a singularity of $o(\frac{1}{r})$ and is therefore *strongly singular*.

2.2 Discretisation

For the solution of Equations (1) we can subdivide the boundary into iso-parametric elements and specify an interpolation inside element e as follows:

$$\mathbf{x}^e = \sum_{i=1}^{I} N_i(\xi)\, \mathbf{x}_i^e \tag{9}$$

$$\mathbf{u}^e = \sum_{i=1}^{I} N_i(\xi)\, \mathbf{u}_i^e \tag{10}$$

$$\mathbf{t}^e = \sum_{i=1}^{I} N_i(\xi)\, \mathbf{t}_i^e \tag{11}$$

where $\mathbf{x}_n^e$,$\mathbf{u}_n^e$ and $\mathbf{t}_n^e$ are nodal values of coordinates, displacements and tractions and N_n are interpolation functions of local coordinate ξ with a range of -1 to 1 and I is the number of nodes. Assuming a quadratic interpolation we have:

$$N_3 = 1 - \xi^2 \tag{12}$$

$$N_1 = \frac{1}{2}(1-\xi) - \frac{1}{2}N_3 \tag{13}$$

$$N_2 = \frac{1}{2}(1-\xi) + \frac{1}{2}N_3 \tag{14}$$

The boundary element and shape functions are shown in Figure 2. The vector tangential to the curve can be computed by:

$$\mathbf{V}_\xi = \frac{\partial \mathbf{x}^e}{\partial \xi} \tag{15}$$

and the length of this vector (*Jacobian*) is:

$$J = \sqrt{v_x^2 + v_y^2} \tag{16}$$

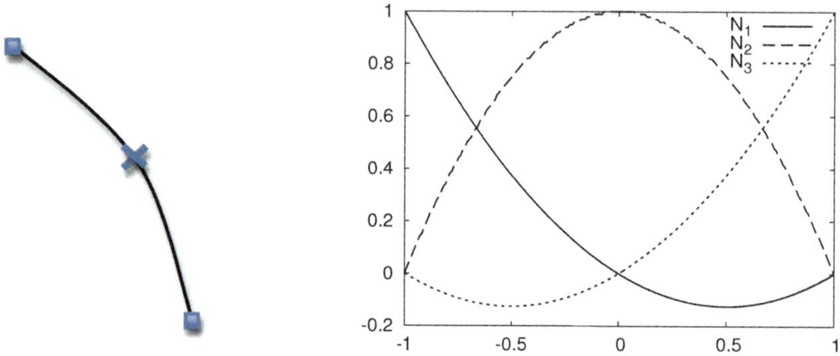

Figure 1. Isoparamteric element and quadratic shape functions

The unit tangential vector is given by

$$\mathbf{v}_\xi = \begin{pmatrix} v_{\xi x} \\ v_{\xi y} \end{pmatrix} = \mathbf{V}_\xi J^{-1} \tag{17}$$

and the vector normal to the curve is given by:

$$\mathbf{n} = \begin{pmatrix} v_{\xi y} \\ -v_{\xi x} \end{pmatrix} \tag{18}$$

The discretized integral equation (1) can now be written as:

$$\begin{aligned} \mathbf{c}\,(P)\ \mathbf{u}\,(P) &= \sum_{e=1}^{E}\sum_{i=1}^{I}\Big(\int_{S_e} \mathbf{U}\,(P,Q)\ N_i\ dS\Big)\ \mathbf{t}_i^e \\ &- \sum_{e=1}^{E}\sum_{i=1}^{I}\Big(\int_{S_e} \mathbf{T}\,(P,Q)\ N_i\ dS\Big)\ \mathbf{u}_i^e \end{aligned} \tag{19}$$

where the summation is over the number of boundary elements E and the number of boundary element nodes I respectively. Using point collocation (i.e. by placing P at the mesh nodes), A matrix equations are obtained, where A is the number of nodes.

$$\begin{aligned}\mathbf{c}(P_a)\ \mathbf{u}(P_a) &= \sum_{e=1}^{E}\sum_{i=1}^{I} \triangle \mathbf{U}_{i,a}^{e}\ \mathbf{t}_n^e \\ &- \sum_{e=1}^{E}\sum_{i=1}^{I} \triangle \mathbf{T}_{i,a}^{e}\ \mathbf{u}_n^e \\ &for \quad a = 1, 2, 3, \ldots, A\end{aligned} \tag{20}$$

where

$$\triangle \mathbf{U}_{i,a}^{e} = \int_{S_e} \mathbf{U}(P_a, Q)\ N_i\, dS \tag{21}$$

$$\triangle \mathbf{T}_{i,a}^{e} = \int_{S_e} \mathbf{T}(P_a, Q)\ N_i\, dS \tag{22}$$

This can be written in matrix form as

$$[\mathbf{T}]\{\mathbf{u}\} = [\mathbf{U}]\{\mathbf{t}\} \tag{23}$$

where the vectors $\{\mathbf{u}\}$ and $\{\mathbf{t}\}$ contain the complete set of displacement and traction components at the nodes of the boundary element mesh. $[\mathbf{T}]$ and $[\mathbf{U}]$ are assembled coefficient matrices, where $[\mathbf{T}]$ includes the free term $\mathbf{c}(P)$.

Computation of internal results Once the solution for the unknowns at the boundary has been obtained the displacements at any point P inside the domain can be obtained by:

$$\mathbf{u}(P) = \int_S \mathbf{U}(P, Q)\,\mathbf{t}(Q)\, dS - \int_S \mathbf{T}(P, Q)\,\mathbf{u}(Q)\, dS \tag{24}$$

From this the strains may be obtained by differentiation. By applying Hooke's law the stresses at point P can be computed by:

$$\boldsymbol{\sigma}(P) = \int_S \mathbf{S}(P, Q)\,\mathbf{t}(Q)\, dS - \int_S \mathbf{R}(P, Q)\,\mathbf{u}(Q)\, dS + \boldsymbol{\sigma}_0 \tag{25}$$

where $\mathbf{S}$ and $\mathbf{R}$ are derived fundamental solutions (see Beer et al. (2008)).

Because of the high order of singularity of the fundamental solutions the computation of stresses very near to or on the boundary is difficult. Therefore we may adopt the method of *stress recovery* (see Beer et al. (2008)) for computing the tangential stress at the boundary.

We first work out the strain along a boundary element by

$$\varepsilon_{\bar{x}} = \frac{\partial u_{\bar{x}}}{\partial \bar{x}} = \frac{\partial(\mathbf{u} \bullet \mathbf{v}_\xi)}{\partial \bar{x}} = \frac{\partial(\mathbf{u} \bullet \mathbf{v}_\xi)}{\partial \xi}\frac{\partial \xi}{\partial \bar{x}} = \frac{\partial(\mathbf{u} \bullet \mathbf{v}_\xi)}{\partial \xi} J^{-1} \tag{26}$$

where

$$\frac{\partial(\mathbf{u} \bullet \mathbf{v}_\xi)}{\partial \xi} = \frac{\partial \mathbf{u}}{\partial \xi} \bullet \mathbf{v}_\xi + \mathbf{u} \bullet \frac{\partial \mathbf{v}_\xi}{\partial \xi} \tag{27}$$

and assuming that J is nearly constant

$$\frac{\partial \mathbf{v}_\xi}{\partial \xi} = \frac{\partial^2 \mathbf{x}}{\partial \xi^2} J^{-1} \tag{28}$$

As this involves a second derivative of the shape function, this term can usually be neglected.

The stress in tangential direction can then be computed using Hooke's law by

$$\sigma_{\bar{x}} = E\,\varepsilon_{\bar{x}} + \nu\, t_{\bar{y}} + \sigma_{\bar{x}0} \tag{29}$$

where $t_{\bar{y}}$ is the traction normal to the boundary and $\sigma_{\bar{x}0}$ is the tangential component of the virgin stress.

2.3 Numerical Evaluation of Integrals

Here we discuss the numerical evaluation of the integrals. We divide the integrals into the following cases:

- Regular integrals: This is the case when the collocation point does not lie on the element e or - in the case it does - the shape function tends to zero as collocation point is approached therefore canceling out the singularity.
- Weakly singular integrals: This is the case when the collocation point lies on the element, the shape function does not tend to zero as the point is approached and involves the integration of the Kernel **U**
- Strongly singular integrals: This is the case when the collocation point lies on the element, the shape function does not tend to zero as the point is approached and involves the integration of the Kernel **T**. This integral only exists as Cauchy principal value.

Regular integration For case 1 we can employ Gauss Quadrature. The regular integrals are given by:

$$\triangle \mathbf{U}^{e}_{i,a} = \int_{-1}^{+1} \mathbf{U}\,(P_a, Q)\ N_i\ J\ d\xi \tag{30}$$

$$= \sum_{g=1}^{G} \mathbf{U}\,(P_a, Q(\xi_g))\ N_i(\xi_g)\ J(\xi_g)\ W_g \tag{31}$$

$$\triangle \mathbf{T}^{e}_{i,a} = \int_{-1}^{+1} \mathbf{T}\,(P_a, Q)\ N_i\ J\ d\xi \tag{32}$$

$$= \sum_{g=1}^{G} \mathbf{T}\,(P_a, Q(\xi_g))\ N_i(\xi_g)\ J(\xi_g)\ W_g \tag{33}$$

In the above G is the number of Gauss points (chosen depending on the proximity of the collocation point to the element), ξ_g are coordinates of Gauss points, J is the Jacobian of the transformation between the local and the global coordinate system and W_g are weights. Care has to be taken when the collocation point is very near to the element as the integrand becomes nearly singular and the number of Gauss points has to be increased considerably. Sometimes it is advantageous to subdivide the integration region.

Weakly singular integrals The kernel $\mathbf{U}$ has a singularity of $o(ln(\frac{1}{r}))$. The product $\mathbf{U}N_i$ tends to infinity for cases where N_i does not tend to zero at the singularity point.

For this we can use a coordinate transformation between the coordinate ξ and a coordinate γ as proposed by Johnston and Elliott (2001), that results in a zero Jacobian at the collocation point, thereby canceling the singularity. This means that the Integration in the γ coordinate space can be carried out using normal Gauss integration.

For example if the collocation point is at $\gamma = -1$ the transformation is given by

$$\xi(\gamma) = \frac{1}{2}(\gamma^2 - 1) + \gamma + 1 \tag{34}$$

The Jacobian of this transformation is $J(\gamma) = 1 + \gamma$ and approaches 0 for $\gamma = -1$.

The Integration can be now carried out using standard Gauss Quadrature

$$\triangle \mathbf{U}^e_{i,a} = \int_{-1}^{1} \mathbf{U}(P_a, Q(\gamma))\, N_i(\xi(\gamma))\, J(\xi) J(\gamma)\, d\gamma \tag{35}$$

$$= \sum_{g=1}^{G} \mathbf{U}(P_a, Q(\gamma_g))\, N_i(\xi(\gamma_g))\, J(\xi_g) J(\gamma_g) W_g \tag{36}$$

Strongly singular integrals The kernel $\mathbf{T}$ has a singularity of $o(\frac{1}{r})$. For the case where the shape function does not tend to zero as the singularity is approached the integral of $\mathbf{T}\,(P_a, Q)\, R_{i,p}$ only exist as Cauchy principal value. However the singularity can be removed by considering that for a finite domain a rigid body motion is not associated with any tractions. Solving Equation (1) for $\mathbf{u}(Q) = \mathbf{I}$ (where $\mathbf{I}$ is a unit vector) and $\mathbf{t}(Q) = 0$ we obtain for the rigid body mode

$$\mathbf{c}\,(P) = -\int_S \mathbf{T}\,(P,Q)\, dS \tag{37}$$

Substitution into Equation (1) gives

$$-\int_S \mathbf{T}\,(P,Q)\, dS\, \mathbf{u}\,(P) = \int_S \mathbf{U}\,(P,Q)\, \mathbf{t}\,(Q)\, dS - \int_S \mathbf{T}\,(P,Q)\, \mathbf{u}\,(Q)\, dS \tag{38}$$

or

$$\int_S \mathbf{U}\,(P,Q)\, \mathbf{t}\,(Q)\, dS = \int_S \mathbf{T}\,(P,Q)\, (\mathbf{u}\,(Q) - \mathbf{u}(P)) dS \tag{39}$$

It can be seen that as Q approaches the collocation point P, the singularity is cancelled out as $\mathbf{T}$ is multiplied by zero.

As pointed out by Beer et al. (2008) the rigid body motion trick may also be applied to infinite domains, by considering a circle of inclusion of radius R and by assuming R to be infinity. In this case we need to add the contribution of the circular boundary (the "azimuthal" integral), i.e.

$$\mathbf{c}\,(P) = -(\int_S \mathbf{T}\,(P,Q)\, dS + \int_{S_\infty} \mathbf{T} dS) \tag{40}$$

The "azimuthal" integral can be computed analytically :

$$\int_{S_\infty} \mathbf{T} dS = -\mathbf{I} \tag{41}$$

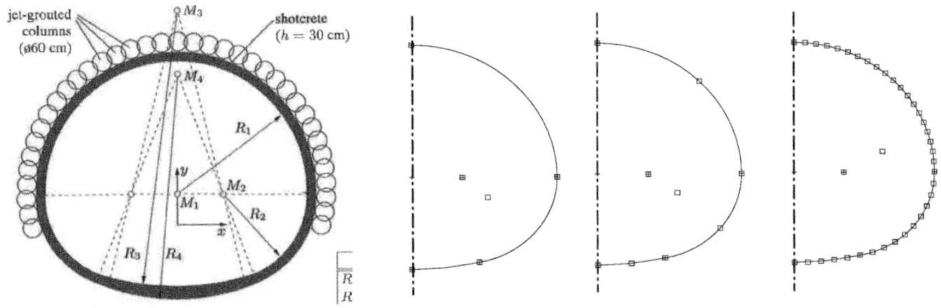

Figure 2. Geometry of NATM tunnel and half of the BEM meshes

2.4 Example

As example we show the simulation of an NATM tunnel. For the purpose of demonstrating the BEM solution it was assumed that the modulus of Elasticity was 10 000 MPa and the vertical virgin stress was -1.0 MPa (compression) and the horizontal virgin stress was zero. Figure 2 shows the geometry of the tunnel and three different meshes with quadratic Boundary Elements. The last mesh can be associated with a converged solution, i.e. the change in the results between two refined meshes was found to be negligible. Figure 3 shows the variation of the resultant displacement along half of the right the tunnel wall starting from the top. Figure 4 shows the variation of the tangential stress computed by Eq.29 divided by the virgin stress (stress concentration factor). The jump in stress between two elements can be seen for the coarse mesh. The convergence properties will be discussed next, together with a comparison of the accuracy with the isogeometric method.

2.5 Conclusion

Here we have presented the conventional Boundary Element Method with Serendipity type shape functions.

We will see next that the introduction of isoparametric concepts will lead to an increase in accuracy and user friendliness.

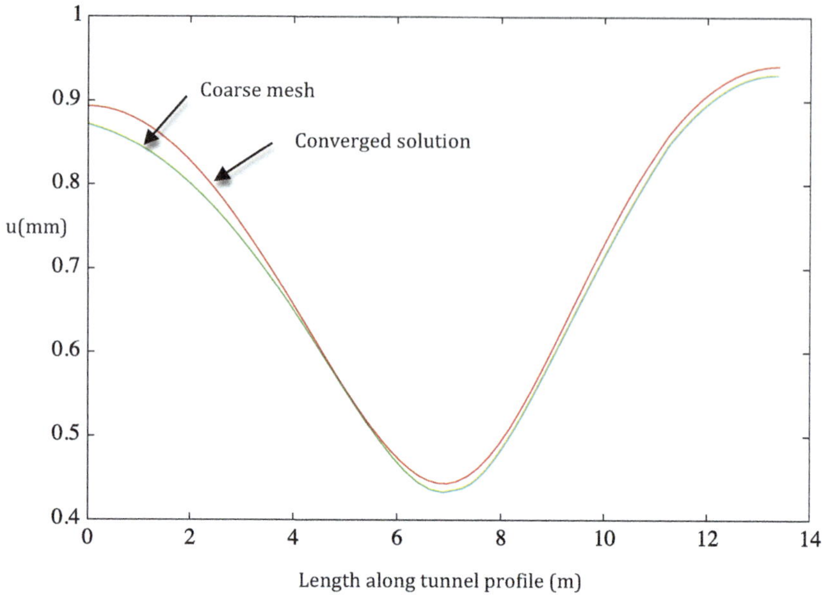

Figure 3. Plot of the resultant displacement along the right half of the tunnel

3 Introduction to isogeometric BEM for plane problems

3.1 Geometry description with NURBS

In the isogeometric concept NURBS basis functions are used instead of the iso-parametric shape functions. Parameters of NURBS patches are specified at control points, which may not be interpolatory, i.e. may not lie on the curve. As an example we compare the geometrical representation of a quarter circle by one parabolic element and one NURBS patch in Figure 5. We see that whereas the control points at the ends coincide with the node points the third control point does not lie on the curve.

For a 2-D problem, the geometrical shape is described by:

$$\mathbf{x} = \sum_{i=1}^{I} R_{i,p}(u)\,\mathbf{B}_i \tag{42}$$

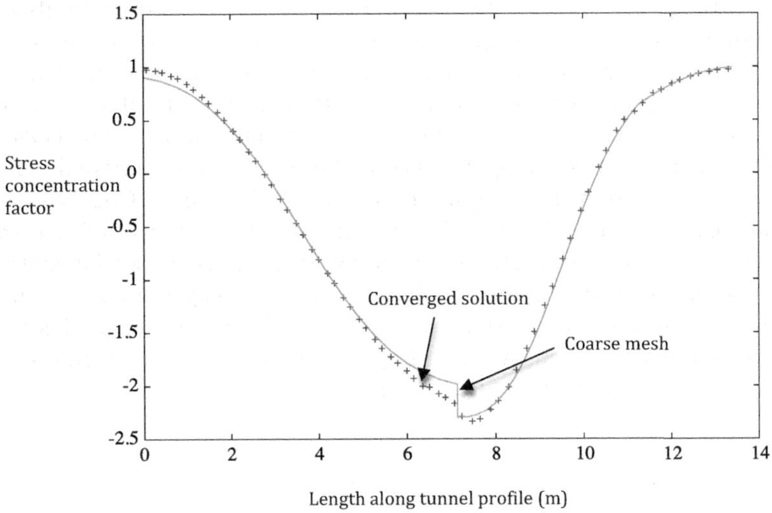

Figure 4. Plot of the tangential stress along right half of tunnel

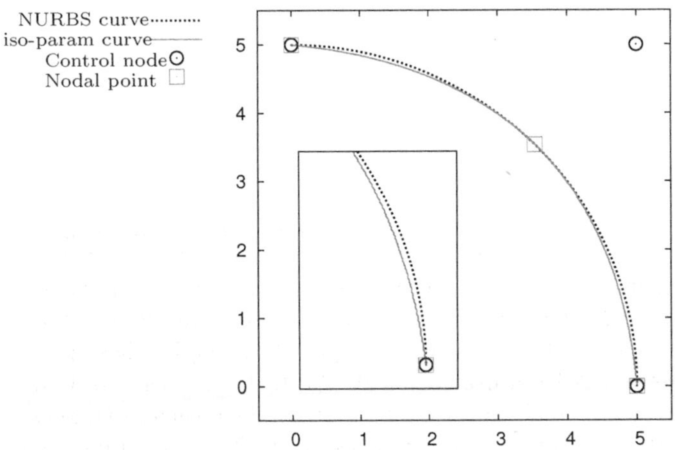

Figure 5. Description of a quarter circle by one quadratic boundary element and one NURBS

where

$$R_{i,p}(u) = \frac{N_{i,p}(u)\, w_i}{\sum_{j=1}^{I} N_{j,p}(u)\, w_j} \tag{43}$$

$N_{i,p}(u)$ are B-spline basis functions of local coordinates u of order p ($p = 0$ constant, $p = 1$ linear, $p = 2$ quadratic etc.). I is the number of control points B_i and w_i are their associated weights, see Piegl and Tiller (1997)

One of the important elements of B-splines is the *knot vector*, defined as a non-decreasing sequence of coordinates in the parameter space u, written $\Xi = \{u_1, u_2, \ldots, u_{I+p+1}\}$. The coordinates u_i of the *knot vector* are called *knots* and the half-open interval $[u_i, u_{i+1})$ is called the *ith knot span*. For example, for the definition of the quarter circle in Figure 5 with a B-spline of order 2, we use an *open knot vector* $\Xi = \{0, 0, 0, 1, 1, 1\}$, where knots are repeated $p+1$ times at the beginning and the end. The knot vector is made of $I + p + 1$ components. The local coordinate u may have any range but here we will restrict them to a range of 0 to 1.

The B-spline basis functions are described for $p = 0$

$$N_{i,0}(u) = \begin{cases} 1 & \text{if} \quad u_i \leqslant u < u_{i+1} \\ 0 & \text{otherwise} \end{cases} \tag{44}$$

Higher order basis functions ($p = 1, 2, 3, \ldots$) are defined by

$$N_{i,p}(u) = \frac{u - u_i}{u_{i+p} - u_i} N_{i,p-1}(u) + \frac{u_{i+p+1} - u}{u_{i+p+1} - u_{i+1}} N_{i+1,p-1}(u) \tag{45}$$

This is a recursive relation requiring the evaluation of lower order basis functions.

The values required to complete the definition of the quarter circle with NURBS, are the weights defined as $w = \begin{pmatrix} 1 & 1/\sqrt{2} & 1 \end{pmatrix}$. In Figure 6 we compare the quadratic Serendipity shape functions and the NURBS basis functions.

We can see that there are subtle differences: The numbering is different and the NURBS basis functions for the corner nodes are not zero at the center. This will need special consideration in the implementation.

For the computation of the outward normal and Jacobian we need the derivative of the NURBS. This is given by

$$\frac{d}{du} R_{i,p}(u) = w_i \, \frac{W(u)\, N'_{i,p}(u) - W'(u)\, N_{i,p}(u)}{W^2(u)} \tag{46}$$

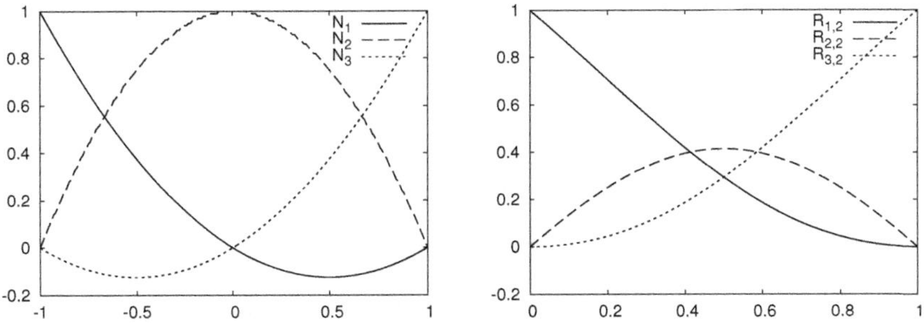

Figure 6. Comparison of quadratic shape functions (left) and NURBS of order 2 (right)

with

$$N'_{i,p}(u) = \frac{d}{du}N_{i,p}(u) = \frac{p}{u_{i+p} - u_i} N_{i,p-1}(u) - \frac{p}{u_{i+p+1} - u_{i+1}} N_{i+1,p-1}(u) \tag{47}$$

$$W(u) = \sum_{j=1}^{I} N_{j,p}(u)\, w_j \tag{48}$$

$$W'(u) = \sum_{j=1}^{I} N'_{j,p}(u)\, w_j \tag{49}$$

Equation 47 is the derivation of the *ith*-B-spline basis function, which is represented in terms of lower order B-spline basis functions. $W(u)$ is called *weighting function* and defined by the sum over all basis functions $N_{i,p}$ multiplied with their weights w_i. Equation 49 defines the derivation of the weighting function. Using this, the outward normal of a NURBS curve is given by

$$\mathbf{n}(u) = \begin{pmatrix} \frac{dy}{du} \\ -\frac{dx}{du} \end{pmatrix} \tag{50}$$

and therefore the Jacobian can be computed by

$$J(u) = \sqrt{n_x^2 + n_y^2} \tag{51}$$

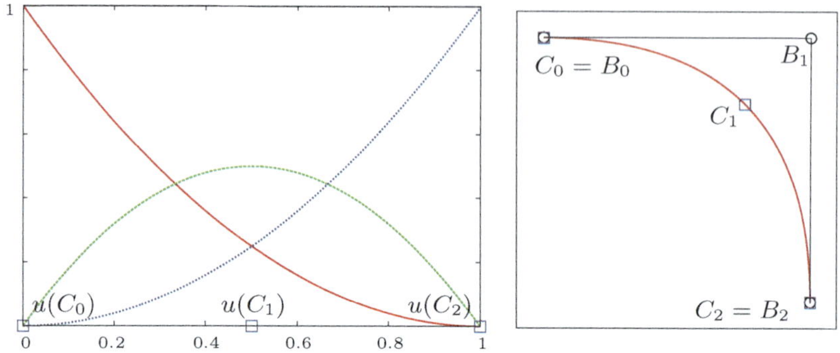

Figure 7. Collocation points C_i in the parametric and physical space.

3.2 Description of the unknowns

In the isogeometric method we use the NURBS also for the variation of the tractions and the displacements:

$$\mathbf{t}^e = \sum_{i=1}^{I} R_{i,p}(u)\, \mathbf{q}_i^e \tag{52}$$

$$\mathbf{u}^e = \sum_{i=1}^{I} R_{i,p}(u)\, \mathbf{d}_i^e \tag{53}$$

where $\mathbf{q}_i^e$ and $\mathbf{d}_i^e$ are the parameters for $\mathbf{t}$ and $\mathbf{u}$ at control point i.

To apply this to the discretisation of the integral equation we have to consider that control points may not lie on the boundary. Therefore, we have to compute the coordinates of collocation points C_i. The local coordinate of the collocation points can be computed using Greville abscissae (see Greville (1964)):

$$u(C_i) = \frac{u_{i+1} + u_{i+2} + \cdots + u_{i+p}}{p} \qquad i = 0, 1, \ldots, I \tag{54}$$

For the discretisation of the quarter circle with one NURBS patch of order $p = 2$, the local coordinates of the collocation points are $\begin{pmatrix} 0 & 0.5 & 1 \end{pmatrix}$ (Figure 7).

The discretised integral equation is now

$$\mathbf{c}\,(P_a)\sum_{i=1}^{I} R_{i,p}(u(P_a))\,\mathbf{d}_i^{ec} = \sum_{e=1}^{E}\sum_{i=1}^{I}\left(\int_{S_e} \mathbf{U}\,(P_a, Q)\;R_{i,p}\,dS\right)\,\mathbf{q}_i^e$$
$$-\sum_{e=1}^{E}\sum_{i=1}^{I}\left(\int_{S_e} \mathbf{T}\,(P_a, Q)\;R_{i,p}\,dS\right)\,\mathbf{d}_i^e \qquad (55)$$
$$for \quad a = 0, 1, 2, \ldots, A$$

where ec indicates the NURBS patch where the collocation point resides and E is the number of NURBS patches. P_a refers to the collocation points, whose coordinates are computed for each NURBS patch using Equation 54. The computation of the free term can be avoided using the rigid body motion trick introduced earlier. For a pure Neuman problem, where tractions are given, the equation can be re-written in matrix form as

$$[\mathbf{T}]\,\{\mathbf{d}\} = \{\mathbf{F}\} \qquad (56)$$

where $\{\mathbf{d}\}$ is a vector of unknown displacement parameters at all control points.

3.3 NUMERICAL EVALUATION OF INTEGRALS

The integration over NURBS is carried out using the same numerical integration method as for the conventional BEM. If we integrate over a NURBS patch attention has to be paid to the fact that shape functions may be zero over parts of the patch. Consider a NURBS patch with a knot vector $\Xi_u = \{0, 0, 0, 0.5, 1, 1, 1\}$. In Figure 8 we see that shape functions 1 and 4 have only a limited range. As with the conventional BEM we divide the integrals into a number of cases.

Regular integration

For case 1 we can employ Gauss Quadrature and this means that we have to transform from NURBS coordinates u to coordinates ξ, that range from -1 to +1. This transformation is given by

$$u = \frac{s_u}{2}\,(\xi + 1) + u_s \qquad (57)$$

where s_u is the size of the subregion and u_s is the local coordinate of the start of the subregion. The Jacobian of this transformation corresponds to

$$\frac{du}{d\xi} = \frac{s_u}{2} \qquad (58)$$

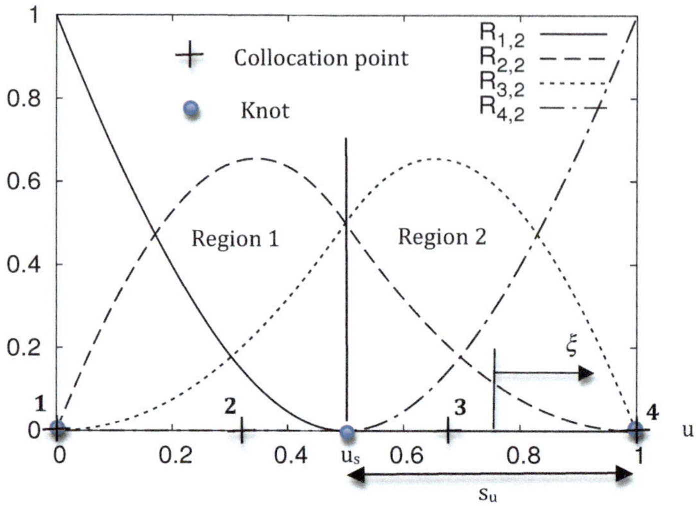

Figure 8. Example of NURBS patch where the integration region has to be split into two

The regular integrals are given by

$$\triangle \mathbf{U}^{e}_{a,i} = \int_{-1}^{+1} \mathbf{U}\left(P_a, Q\right) \, R_{i,p} \, J \, \frac{du}{d\xi} \, d\xi \tag{59}$$

$$= \sum_{g=1}^{G} \mathbf{U}\left(P_a, Q(\xi_g)\right) \, R_{i,p}(\xi_g) \, J(\xi_g) \, \frac{du}{d\xi} \, W_g \tag{60}$$

$$\triangle \mathbf{T}^{e}_{a,i} = \int_{-1}^{+1} \mathbf{T}\left(P_a, Q\right) \, R_{i,p} \, J \, \frac{du}{d\xi} \, d\xi \tag{61}$$

$$= \sum_{g=1}^{G} \mathbf{T}\left(P_a, Q(\xi_g)\right) \, R_{i,p}(\xi_g) \, J(\xi_g) \, \frac{du}{d\xi} \, W_g \tag{62}$$

In the above, G is the number of Gauss points (chosen depending on the proximity of the collocation point to the element), ξ_g are coordinates of Gauss points, J is the Jacobian of the transformation between the NURBS and the global coordinate system and W_g are weights. Care has to be taken when the collocation point is very near to the element as the number of Gauss points has to be increased. Sometimes it is advantageous to further

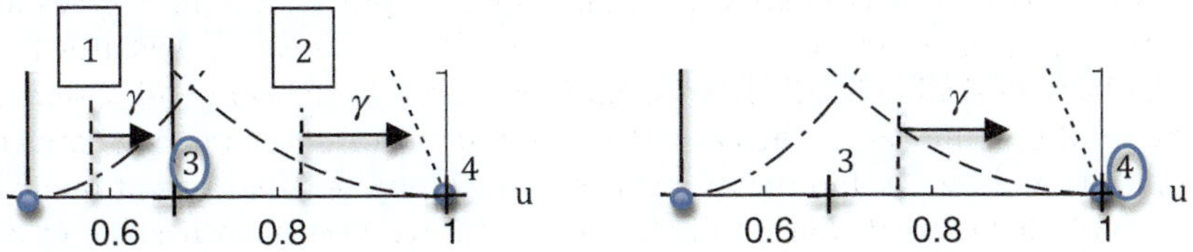

Figure 9. Explanation of the coordinate transformation between γ and u, left for an arbitrary position of the collocation point, right where the collocation point is at an edge

subdivide the integration region.

Weakly singular integrals

Kernel $\mathbf{U}$ has a singularity of $o(ln(\frac{1}{r}))$. The product $\mathbf{U}R_{i,p}$ tends to infinity for cases where $R_{i,p}$ does not tend to zero at the singularity point.

For this we can use a coordinate transformation between the NURBS coordinate u and a coordinate γ with the limits -1 and +1 as introduced previously, that results in a zero Jacobian at the collocation point, thereby canceling the singularity. This means that the Integration in the γ coordinate space can be carried out using normal Gauss integration.

If the collocation point is at an arbitrary position u_i inside the integration region, it has to be subdivided into 2 subregions as shown in Figure 17. The transformation for sub-region 1 is given by

$$u(\gamma) = u_i \, \frac{s_u}{2} \left(\frac{1}{2}(1-\gamma^2) + \gamma + 1 \right) + u_s \tag{63}$$

The Jacobian of this transformation is $J(\gamma) = u_i \, \frac{s_u}{2} \, (1-\gamma)$ and tends to zero for $\gamma = 1$. For subregion 2 the transformation is

$$u = \frac{s_u - u_i}{2} \left(\frac{1}{2}(\gamma^2 - 1) + \gamma + 1 \right) + u_i + u_s \tag{64}$$

The Jacobain of this transformation is $J(\gamma) = \frac{s_u - u_i}{2} \, (\gamma + 1)$ and tends to zero for $\gamma = -1$.

For the case whn the collocation points are on the edges of the element, the coordinate transformation is given by for the collocation at $\gamma = -1$

$$u(\gamma) = \frac{s_u}{2} \left(\frac{1}{2}(\gamma^2 - 1) + \gamma + 1 \right) + u_s \tag{65}$$

the Jacobian of this transformation is $J(\gamma) = \frac{s_u}{2}(1+\gamma)$ and approaches 0 for $\gamma = -1$. For the collocation point at $\gamma = 1$ we have

$$u(\gamma) = \frac{s_u}{2}\left(\frac{1}{2}(1-\gamma^2) + \gamma + 1\right) + u_s \tag{66}$$

the Jacobian of this transformation is $J(\gamma) = \frac{s_u}{2}(1-\gamma)$ and approaches 0 for $\gamma = 1$.

The Integration within the subregion can be now carried out using standard Gauss Quadrature

$$\triangle \mathbf{U}^e_{a,i} = \int_{-1}^{1} \mathbf{U}(P_a, Q(\gamma))\, R_{i,p}(u(\gamma))\, J(u)J(\gamma)\, d\gamma \tag{67}$$

$$= \sum_{g=1}^{G} \mathbf{U}(P_a, Q(\gamma_g))\, R_{i,p}(u(\gamma_g))\, J(u_g)J(\gamma_g)W_g \tag{68}$$

Strongly singular integrals

The singularity can be removed by considering a rigid body motion as explained in the previous section.

3.4 Example

As example we choose the same as for the iso-parametric BEM. Figure 10 shows the geometry definition used for the analysis. It consists of 6 NURBS patches of order $p = 2$. Since the design cross-section of the tunnel is described by arcs, this discretisation represents the geometry exactly. The number of degrees of freedom is equal to the one for the coarsest iso-parametric mesh. The results are presented in Figures 11 for the displacements and 12 for the stresses and compared with the isoparametric analysis. It can be seen that for the same number of degrees of freedom the results of the isogeometric BEM are much more accurate than those of the iso-parametric BEM. The reason for this is that the geometry is described exactly and that the basis functions used for approximating the displacements provide a more accurate description of the actual situation.

3.5 Conclusions

In this section we have introduced the isogeometric concepts into the BEM. We observe two advantages: The geometry description for the tunnel is exact and does not have to be approximated by a mesh. Results of an analysis with few degrees of freedom are much better than for a comparable iso-parametric analysis. Next we look at strategies for refining the solution.

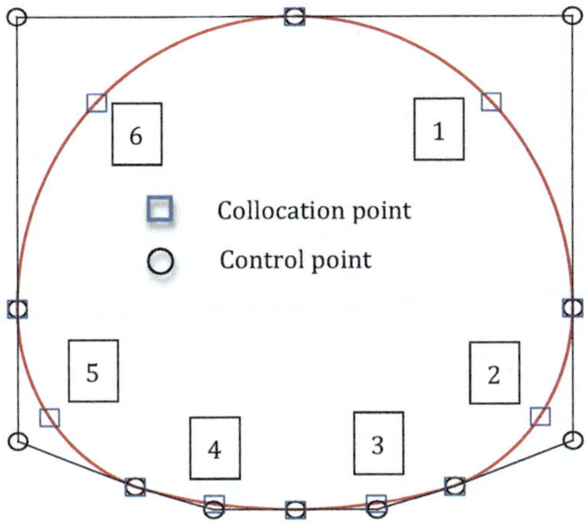

Figure 10. Mesh used for the analysis consisting of 6 NURBS patches

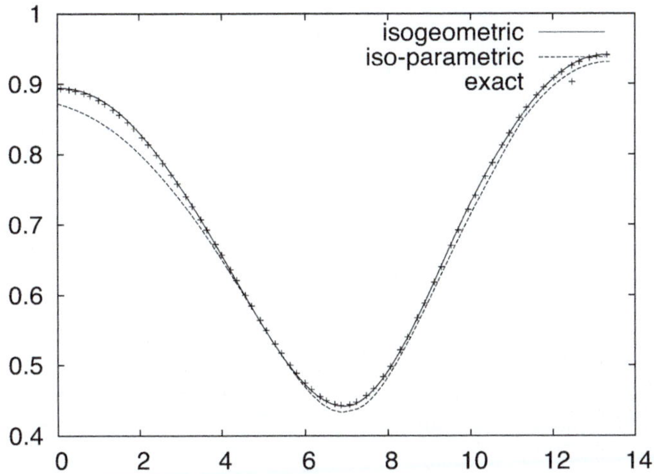

Figure 11. Variation of the resultant displacement along the right half of the tunnel starting from the top

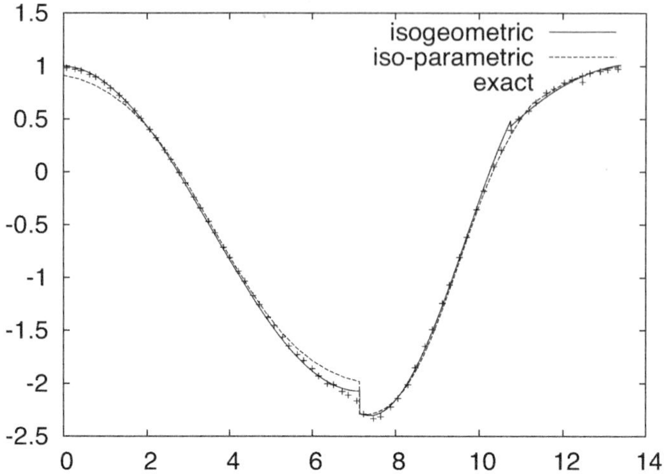

Figure 12. Variation of the stress concentration factor along the right half of the tunnel starting from the top

4 Isogeometric refinement strategies

Here we investigate different methods used in isogeometric analysis for the refinement of the description of the unknown, which is given by:

$$\mathbf{u}^e = \sum_{i=1}^{I_u} R_{i,p_u}(u)\, \mathbf{d}_i^e \tag{69}$$

where R_{i,p_u} are NURBS basis functions of order p_u.

For the refinement we can use either *knot insertion* or *order elevation.* Both expand a given knot vector Ξ, for the geometry description to an *extended knot vector* Ξ_u, such that $\Xi \subset \Xi_u$.

4.1 Knot insertion

For the first stage of *knot insertion* we insert a knot value into the knot vector. We keep the order the same as for the geometry description, i.e. $p_u = p$ but change the original Knot vector ($\Xi_u = \{0, 0, 0, 1, 1, 1\}$) to $\Xi_u = \{0, 0, 0, 0.5, 1, 1, 1\}$. As a consequence, the number of control points increases to $I_u = I + 1$.

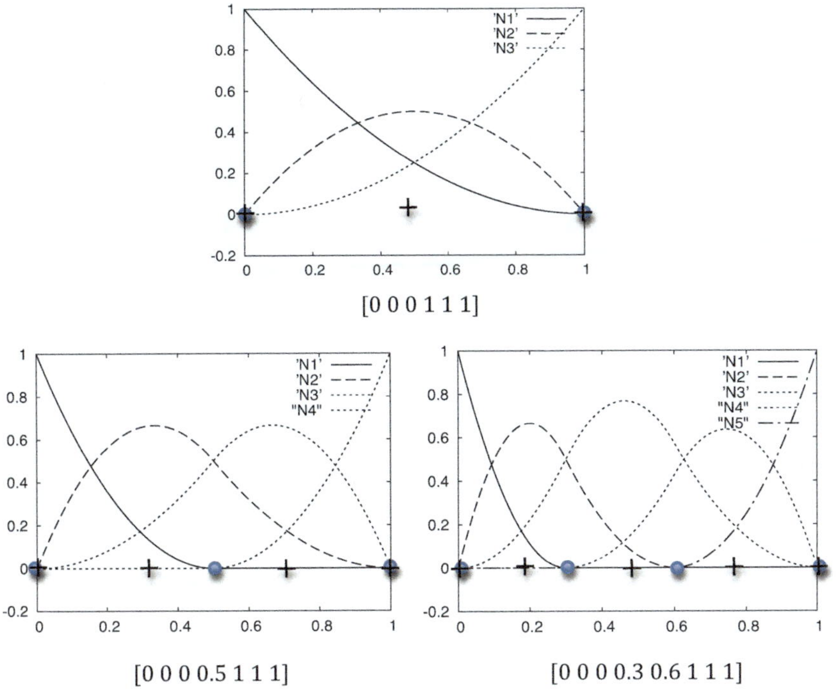

Figure 13. Original basis functions and two stages of refinement by knot insertion for p=2 with the corresponding knot vectors. Circles depict knots, crosses depict collocation points

The new weights w_{iu} corresponding to the resulting four NURBS basis functions, shown in Figure 13 on the left, are a linear combination of the original weights w_i of the geometry description.

$$w_{iu} = \alpha_i \, w_i + (1 - \alpha_i) \, w_{i-1} \qquad \alpha_i = \begin{cases} 1 & i \leqslant k - p \\ \frac{\bar{u} - u_i}{u_{i+p} - u_i} & k - p + 1 \leqslant i \leqslant k \\ 0 & i \geq k + 1 \end{cases} \tag{70}$$

where $\bar{u}$ is the value of the inserted knot, which lies in the knot span $[u_k, u_{k+1})$. Using the weights of the quarter circle $w = \begin{pmatrix} 1 & 1/\sqrt{2} & 1 \end{pmatrix}$ the new weights become $w_{iu} = \begin{pmatrix} 1 & 0.8536 & 0.8536 & 1 \end{pmatrix}$. However it can be seen that the first and last basis function have only local support, i.e. they have zero value over half of the element. This has to be considered when

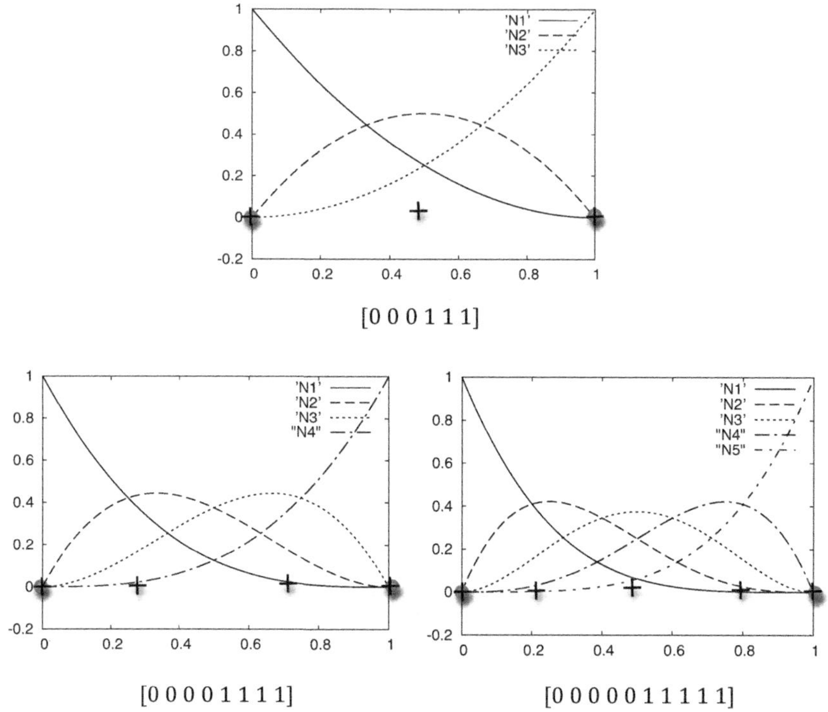

Figure 14. Original basis functions and two stages of order elevation from p=2 (p=3 and p=4) with the corresponding Knot vectors, circles depict Knots, crosses depict collocation points

performing the numerical integration. Since there are now more control points and therefore unknowns we need more equations and this means more collocation points.

For this we use the Greville formula introduced previously, which gives the following result: $u(C_1) = 0, u(C_2) = \frac{1}{3}, u(C_3) = \frac{2}{3}, u(C_4) = 1$.

We can further refine the solution by inserting another knot ($\Xi_u = \{0, 0, 0, 0.3, 0.6, 1, 1, 1\}$ as shown in Figure 13.

4.2 Order elevation

Another possibility is to use *order elevation* for the refinement. This mechanism elevates the order of the NURBS by replicating existing knots.

To get cubic basis functions out of quadratic basis functions, i.e. $p_u = p+1$, the existing knots are replicated 1 more time, which leads to a corresponding knot vector of $\Xi_u = \{0,0,0,0,1,1,1,1\}$. The basis functions are shown in Figure 14. The number of control points is also increased for this case.

For the computation of the new weights w_{iu} it is proposed by Piegl and Tiller to perform the order elevation in the following steps:

1. Subdivide the NURBS into Bézier segments by knot insertion. Thus the basis functions becomes interpolatory at every knot value u_i.
2. Order elevate these Bézier segments.
3. Combine them afterwards to get one single NURBS again.

In the case of the quarter circle the NURBS is already a Bézier segment. Therefore, the new weights w_{iu} are obtained from all original weights w_i by a linear combination.

$$w_{iu} = (1-\alpha_i)\, w_i + \alpha_i\, w_{i-1} \qquad \alpha_i = \begin{cases} 0 & i=1 \\ \frac{i-1}{I} & 1 < i \leqslant I \\ 1 & i = I+1 \end{cases} \tag{71}$$

Using again the weights of the quarter circle $w = \begin{pmatrix} 1 & 1/\sqrt{2} & 1 \end{pmatrix}$ the new weights become $w_{iu} = \begin{pmatrix} 1 & 0.8047 & 0.8047 & 1 \end{pmatrix}$.

The local coordinates of the collocation points are computed for the first order elevation as $u(C_1) = 0, u(C_2) = \frac{1}{4}, u(C_3) = \frac{3}{4}, u(C_4) = 1$ (see Figure 14).

4.3 Geometry independent field approximation

As discussed in the previous section NURBS curves can exactly represent geometry of a problem. Hence the geometry description does not need to be improved. But in general, the NURBS basis functions that are used representing the geometry, may be not sufficient for the description of the solution.

Therefore we propose to use a geometry independent concept for the refinement, i.e. to use different functions for the description of the geometry, the tractions and the displacements. For the moment let us restrict ourselves here to an excavation problem, with known traction values at the boundary. Therefore the unknown are the displacements.

Our approach is to only enrich the solution space using extended NURBS basis functions, i.e. we change the description of the displacements to:

$$\mathbf{u}^e = \sum_{i=1}^{I_u} R_{i,p_u}(u)\, \mathbf{d}_i^e \tag{72}$$

where R_{i,p_u} are NURBS basis functions of order p_u, defined by *extended knot vector* Ξ_u based on the knot vector Ξ of the geometry representation. I_u is the new number of basis functions.

For an excavation problem the traction t_0 can be directly computed from the virgin stress σ_0 and the outward normal $\mathbf{n}$ as

$$t_{x0} = n_x \sigma_{x0} \tag{73}$$

$$t_{y0} = n_y \sigma_{y0} \tag{74}$$

and therefore the discretised equation can be written as

$$\begin{aligned} &\sum_{e=1}^{E} \sum_{i=1}^{I_u} \left(\int_{S_e} \mathbf{T}\,(P_a, Q)\; R_{i,p_u}\, dS \,\right) \mathbf{d}_a^e \\ &= \sum_{e=1}^{E} \left(\int_{S_e} \mathbf{U}\,(P_a, Q)\; t_0(Q)\, dS \,\right) - \mathbf{I} \\ &\qquad for \quad a = 1, 2, 3 ... A_u = I_u \end{aligned} \tag{75}$$

or in matrix form

$$[\mathbf{T}]\,\{\mathbf{d}\} = \{\mathbf{F}\} \tag{76}$$

where A_u is the (new) number of collocation points and $\{\mathbf{F}\}$ is a vector of known values computed from the right hand side of Equation 75.

4.4 Example

As an example we analyze two adjacent NATM tunnels.The ground is subjected to a vertical compressive virgin stress σ_y of magnitude 1.0 MPa and a horizontal virgin stress σ_x of magnitude 0.25 MPa. The elastic material properties are $\nu = 0.25$, $E = 10000$ MPa. This shape can be described exactly by 12 NURBS patches of order 2 (see figure 15). The reference solution is calculated using a conventional iso-parametric analysis with an extremely fine mesh (240 quadratic elements).

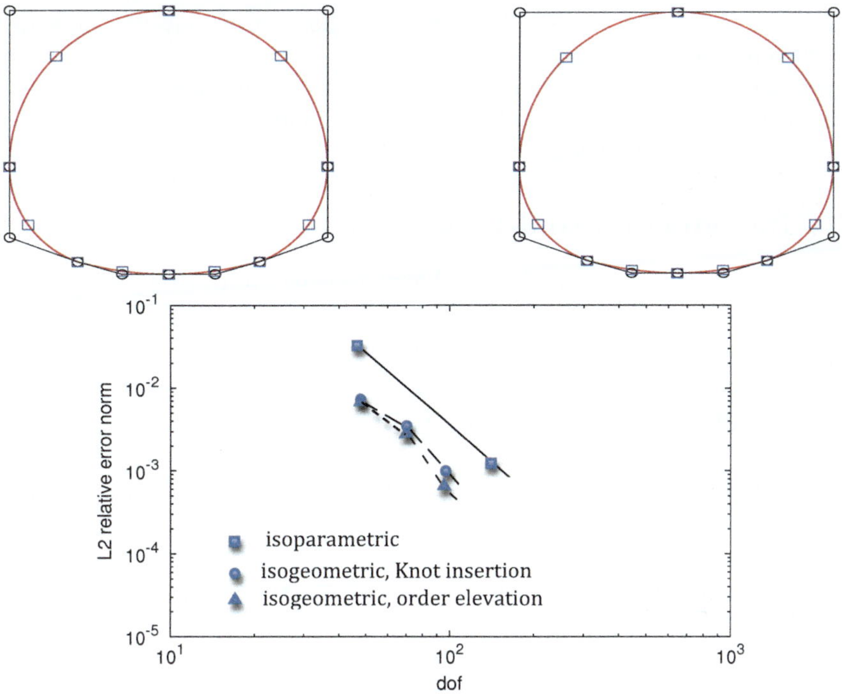

Figure 15. Top: Definition of geometry of NATM tunnel by 12 NURBS showing the control points (circles) and the computed collocation points (squares). Bottom: Convergence of solution

We show the convergence of the solution as a plot of the L^2 norm of the error in the displacement along the boundary versus the degrees of freedom for refinement by Knot insertion and order elevation. Comparison is made with the isoparametric method using h-refinement (i.e increasing the number of elements) and quadratic shape function. It can be seen that the results of the isogeometric methods start with a much lower error and that the convergence rates are equal or better than for the isoparametric analysis with best convergence achieved with a higher order.

4.5 Summary

In this section we have introduced the methods that can be used to refine the solution. Two methods are available: Knot insertion and order elevation. The first method is similar to the h-refinement in conventional analysis but there is a subtle difference: Instead of generating more elements we generate more functions, keeping the integration region unchanged. Therefore the method has the potential for developing simulation methods that do not need any mesh refinement.

5 Introduction to 3-D problems

Here we extend the method to problems in three-dimensional space.

5.1 Boundary integral equation

The boundary integral equation for an elastic continuum without body forces can be written as (see Beer et al. (2008))

$$\mathbf{c}\,(P)\,\mathbf{u}\,(P) \;=\; \int_S \mathbf{U}\,(P,Q)\,\mathbf{t}\,(Q)\,dS - \int_S \mathbf{T}\,(P,Q)\,\mathbf{u}\,(Q)\,dS \quad (77)$$

where P (coordinates x_P, y_P, z_P) is the source point and Q (coordinates x_Q, y_Q, z_Q) is the field point on the boundary S. The coefficient $\mathbf{c}\,(P)$ is a free term related to the boundary geometry.

$\mathbf{u}\,(Q)$ and $\mathbf{t}\,(Q)$ are the displacements and tractions on the boundary and $\mathbf{U}\,(P,Q)$ and $\mathbf{T}\,(P,Q)$ are matrices containing Kelvin's fundamental solutions (*Kernels*) for the displacements and tractions respectively. For a 3-D problem we have

$$\mathbf{U}\,(P,Q) = C * \frac{1}{r}\,(C_1\mathbf{I} + \mathbf{R}) \quad (78)$$

where where $\mathbf{I}$ is the unit matrix, C and C_1 are material constants, r is the distance between point P and Q and

$$\mathbf{R} = \begin{pmatrix} r_x^2 & r_x r_y & r_x r_z \\ r_y r_x & r_y^2 & r_y r_z \\ r_z r_x & r_z r_y & r_z^2 \end{pmatrix} \quad (79)$$

where

$$\begin{aligned} r_x &= \frac{1}{r}\,(x_Q - x_P) \\ r_y &= \frac{1}{r}\,(y_Q - y_P) \\ r_z &= \frac{1}{r}\,(z_Q - z_P) \end{aligned} \quad (80)$$

and r is the distance between P and Q. As expected $\mathbf{U}$ is symmetric.

The fundamental solution for the tractions is given by

$$\mathbf{T}\,(P,Q) = C_2\,\frac{1}{r^2}((C_3\mathbf{I} + 3\,\mathbf{R}) * \cos\theta + C_3\mathbf{R}_1) \tag{81}$$

where

$$\cos\theta = n_x r_x + n_y r_y + n_z r_z \tag{82}$$

is the angle between the position vector $\mathbf{r}$ and the vector normal to the boundary $\mathbf{n}$ (components n_x, n_y, n_z), furthermore

$$\mathbf{R}_1 = \begin{pmatrix} 0 & n_x r_y - n_y r_x & n_x r_z - n_z r_x \\ n_y r_x - n_x r_y & 0 & n_y r_z - n_z r_y \\ n_z r_x - n_x r_z & n_z r_y - n_y r_z & 0 \end{pmatrix} \tag{83}$$

$\mathbf{T}$ can be split into a symmetric and antisymmetric part.

$$\mathbf{T} = \frac{1}{r^2}(\mathbf{T}_s + \mathbf{T}_a) \tag{84}$$

where

$$\begin{aligned} \mathbf{T}^s &= C_2\,(C_3\mathbf{I} + 3\,\mathbf{R} * \cos\theta) \\ \mathbf{T}^a &= C_2\;C_3\;\mathbf{R}_1 \end{aligned} \tag{85}$$

5.2 Geometry description with NURBS

For a 3-D problem the geometrical shape is described by a tensor product of NURBS:

$$\mathbf{x} = \sum_{b=1}^{B}\sum_{a=1}^{A} R_{a,b}^{p,q}(u,v)\mathbf{x}_{a,b} \tag{86}$$

where $\mathbf{x}_{a,b}$ are the coordinates of the control points, u and v are local coordinates, p and q are the function orders in u and v direction, A and B are the number of control points in u and v direction and

$$R_{a,b}^{p,q} = \frac{N_{a,p}(u)N_{b,q}(v)w_{a,b}}{\sum_{\bar{b}=1}^{B}\sum_{\bar{a}=1}^{A} N_{\bar{a},p}(u)N_{\bar{b},q}(v)w_{\bar{a},\bar{b}}} \tag{87}$$

$N_{a,p}(u)$ and $N_{b,q}(v)$ are B-spline functions of local coordinates u or v of order p or q (0 constant, 1 linear, 2 quadratic etc.) and $w_{a,b}$ are weights.

The B-spline functions are for $p = 0$

$$N_{a,0}(u) = \begin{cases} 1 & \text{if} \quad u_a \leqslant u < u_{a+1} \\ 0 & \text{otherwise} \end{cases} \tag{88}$$

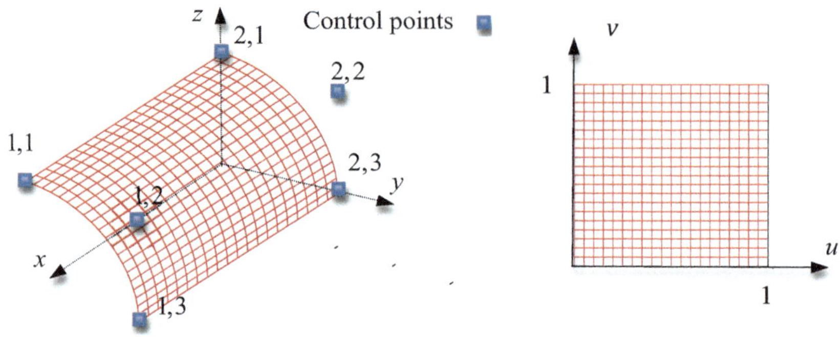

Figure 16. A quarter of a cylinder in global and local coordinate space

and for p=1,2,3

$$N_{a,p}(u) = \frac{u - u_p}{u_{a+p} - u_a} * N_{a,p-1} + \frac{u_{a+p+1} - u}{u_{a+p+1} - u_{a+1}} * N_{a+1,p-1} \tag{89}$$

Similarly for $q = 0$

$$N_{b,0}(v) = \begin{cases} 1 & \text{if} \quad v_b \leqslant v < v_{b+1} \\ 0 & \text{otherwise} \end{cases} \tag{90}$$

and for q=1,2,3

$$N_{b,q}(v) = \frac{v - v_q}{v_{b+q} - v_b} * N_{b,q-1} + \frac{v_{b+q+1} - v}{v_{b+q+1} - v_{b+1}} * N_{b+1,q-1} \tag{91}$$

This is a recursive relation requiring the evaluation of lower order basis functions. u_i (or v_i) are coordinates of a *knot vector*, which is defined as a non-decreasing sequence of coordinates in the parameter space u or v.

For example, for the definition of the quarter cylinder in Figure 16 with NURBS of order $p = 1$ in u- and of order $q = 2$ in v-direction we use the following knot vectors

$$\begin{aligned} \Xi_u &= \begin{pmatrix} 0 & 0 & 1 & 1 \end{pmatrix} \\ \Xi_v &= \begin{pmatrix} 0 & 0 & 0 & 1 & 1 & 1 \end{pmatrix} \end{aligned} \tag{92}$$

where for an *open knot vector*, knots are repeated $p+1$ or $q+1$ times at the beginning and the end. The knot vector is made of $A+p+1$ or $B+q+1$ components. The local coordinates u,v may have any range, but in this paper we will restrict them to a range of 0 to 1.

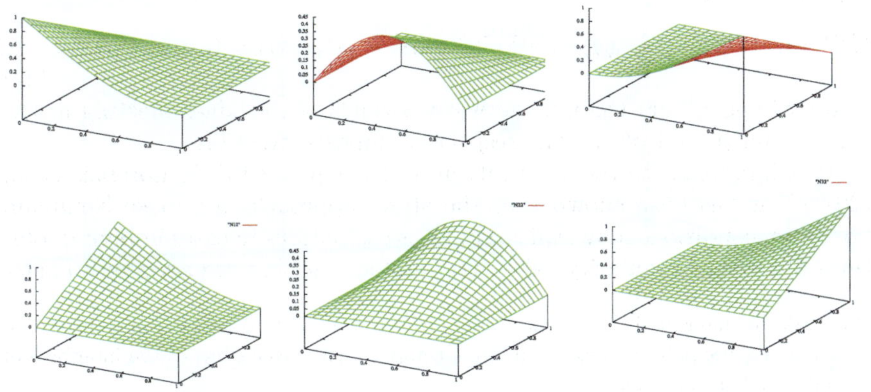

Figure 17. The 6 functions for the geometrical description of the quarter cylinder

The values, required to complete the definition are the weights and control point coordinates, which are given by

$$\mathbf{w} = \begin{pmatrix} 1 & 0.707 & 1 \\ 1 & 0.707 & 1 \end{pmatrix} \tag{93}$$

$$\mathbf{x} = \begin{pmatrix} (2,0,1) & (2,1,1) & (2,1,0) \\ (0,0,1) & (0,1,1) & (0,1,2) \end{pmatrix} \tag{94}$$

The 6 functions for describing a quarter cylinder are shown in Figure 17. Although the functions look similar to the Serendipity functions introduced earlier there are subtle differences: The values of functions are not unity at one point and zero at others and they do not have negative values. This description of the 1/4 cylinder with one NURBS patch is exact.

Degenerate NURBS patches Usually NURBS patches have 4 sides but sometimes it is convenient to use three-sided patches, for example to model a tunnel face. This can be achieved by giving the control points on one side the same coordinates. Figure 18 shows an example of a degenerate patch.

The control point coordinates are given by

$$\mathbf{x} = \begin{pmatrix} (0,0,1) & (0,1,1) & (0,1,0) \\ (0,0,0) & (0,0,0) & (0,0,0) \end{pmatrix} \tag{95}$$

As the location of the coincident control points is approached the Jacobian tends to zero and the outward normal can not be computed.

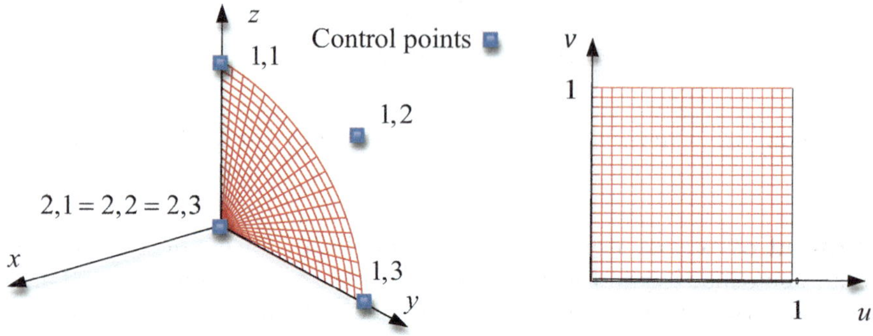

Figure 18. A degenerate patch in global and local coordinate space

However, since these are never evaluated there but at the Gauss points there is no problem. Indeed this situation is beneficial for the evaluation of nearly singular integrands as will be explained later.

5.3 Computation of outward normal and Jacobian

For the computation of the outward normal and the Jacobian we need the derivatives of the NURBS. The derivative to u, for example, is given by

$$\frac{d}{du}R^{p,q}_{a,b}(u,v) = \frac{N' w_{a,b} - W'(u) R^{p,q}_{a,b}(u,v)}{W(u)} \tag{96}$$

with

$$W(u) = \sum_{\hat{a}=1}^{A} \sum_{\hat{b}=1}^{B} N_{\hat{a},p}(u) N_{\hat{b},q}(v) w_{\hat{a},\hat{b}} \tag{97}$$

$$W'(u) = \sum_{\hat{a}=1}^{A} \sum_{\hat{b}=1}^{B} N'_{\hat{a},p}(u) N_{\hat{b},q}(v) w_{\hat{a},\hat{b}} \tag{98}$$

and

$$N' = \frac{d}{du}(N_{a,p}(u,v)) N_{b,q}(u,v) \tag{99}$$

The vector normal to the surface is computed as a vector x-product of the two vectors tangential to the surface. The tangential vector in u-direction is given by

$$\mathbf{V}_u = \frac{\partial}{\partial u}\mathbf{x} = \sum_{b=1}^{B}\sum_{a=1}^{A}(\frac{\partial}{\partial u}R^{p,q}_{a,b}(u,v))\mathbf{x}_{a,b} \tag{100}$$

and the tangential vector in v-direction

$$\mathbf{V}_v = \frac{\partial}{\partial v}\mathbf{x} = \sum_{b=1}^{B}\sum_{a=1}^{A}(\frac{\partial}{\partial v}R^{p,q}_{a,b}(u,v))\mathbf{x}_{a,b} \tag{101}$$

The vector normal to the surface is given by

$$\mathbf{N} = \mathbf{V}_u \times \mathbf{V}_v \tag{102}$$

and the Jacobian is

$$J = |\mathbf{N}| = \sqrt{N_x^2 + N_y^2 + N_z^2} \tag{103}$$

The unit vector in a normal direction is given by:

$$\mathbf{n} = \mathbf{N}\, J^{-1} \tag{104}$$

5.4 NURBS patches of infinite extent

Another situation that occurs in geomechanics is surfaces that extend to infinity. For example, in the case of a very long tunnel, where we are only interested in what happens at the tunnel face. The geometrical shape of an infinite patch can be described by

$$\mathbf{x} = \sum_{b=1}^{B}\sum_{a=1}^{2} R^{\infty}_{a,b}\mathbf{x}_{a,b} \tag{105}$$

with

$$R^{\infty}_{a,b} = M^{\infty}_{a}(u)R^{q}_{b}(v) \tag{106}$$

where M^{∞}_{b} are special shape functions (see Beer and Watson (1989)) that tend to infinity as u approaches 1:

$$M^{\infty}_{1}(u) = (1-2u)/(1-u) \tag{107}$$
$$M^{\infty}_{2}(u) = u/(1-u) \tag{108}$$

The functions are shown in Figure 19.

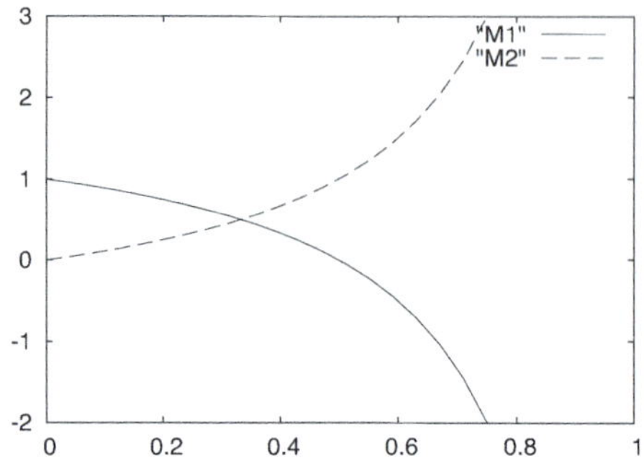

Figure 19. Infinite shape functions

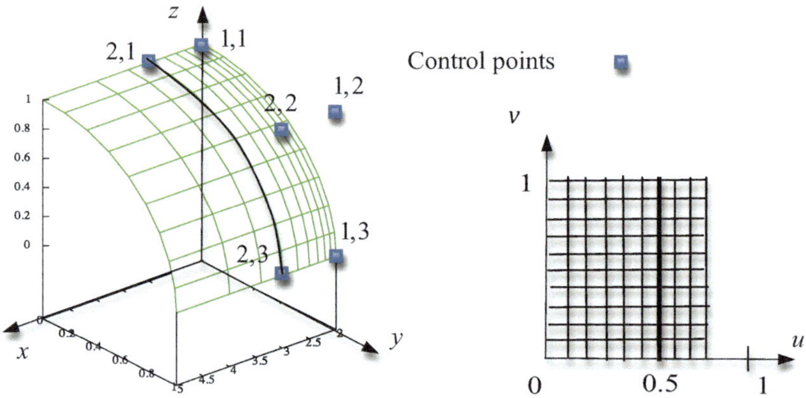

Figure 20. Example of an infinite NURBS patch in global and local coordinate space

Figure 20 shows a NURBS patch that describes a quarter cylinder of infinite extent.

The control point coordinates of the patch are given by

$$\mathbf{x} = \begin{pmatrix} (2,0,1) & (2,1,1) & (2,1,0) \\ (3,0,0) & (3,1,1) & (3,1,0) \end{pmatrix} \tag{109}$$

The derivatives of the function are given by

$$\frac{\partial}{\partial u} R^{\infty}_{a,b}(u,v) = \frac{\partial M^{\infty}_{a}(u)}{\partial u} R^{q}_{b}(v) \tag{110}$$

$$\frac{\partial}{\partial v} R^{\infty}_{a,b}(u,v) = \frac{\partial R^{q}_{b}(v)}{\partial v} M^{\infty}_{a}(u) \tag{111}$$

where the derivatives of $R^q_b(v)$ have been presented previously and

$$\frac{\partial M^{\infty}_{1}}{\partial u} = \frac{-1}{2(1-u^2)} \tag{112}$$

$$\frac{\partial M^{\infty}_{2}}{\partial u} = \frac{1}{2(1-u^2)} \tag{113}$$

Description of the variables In the iso-geometric method we use NURBS also for the variation of the tractions and the displacements:

$$\mathbf{t} = \sum_{b=1}^{B} \sum_{a=1}^{A} R^{p,q}_{a,b}(u,v) \mathbf{q}^{e}_{a,b} \tag{114}$$

$$\mathbf{u} = \sum_{b=1}^{B} \sum_{a=1}^{A} R^{p,q}_{a,b}(u,v) \mathbf{d}^{e}_{a,b} \tag{115}$$

where $\mathbf{q}^{e}_{a,b}$ and $\mathbf{d}^{e}_{a,b}$ are the parameters for $\mathbf{t}$ and $\mathbf{u}$ at control point a, b.

For an infinite patch two assumptions of the variation of $\mathbf{u}$ may be considered.

- *Plane Strain*: Displacements do not vary to infinity

$$\mathbf{u}(u,v) = \mathbf{u}(u=0,v) = \sum_{b=1}^{B} R^{q}_{b}(v) \mathbf{q}^{e}_{b} \tag{116}$$

- *Decay*: Displacements approach zero as infinity is approached.

$$\mathbf{u}(u,v) = (1-u)\,\mathbf{u}(u=0,v) = (1-u) \sum_{b=1}^{B} R^{q}_{b}(v) \mathbf{q}^{e}_{b} \tag{117}$$

It can be shown (see W. Moser and Beer (2004)) that this corresponds to a decay of $O(\frac{1}{r})$.

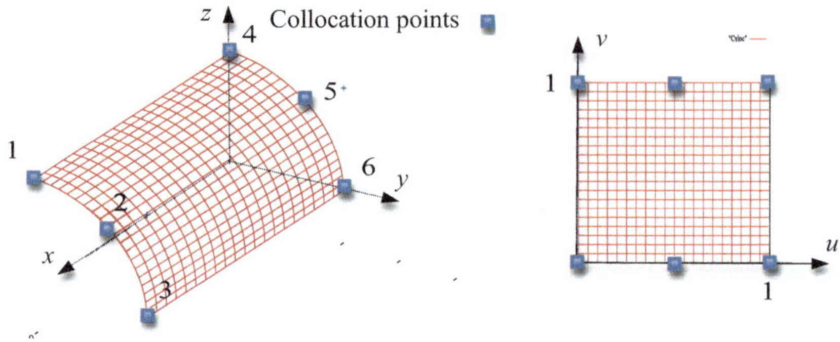

Figure 21. Collocation points for the quarter cylinder

5.5 Implementation

To obtain a system of equations we have to compute the coordinates of the collocation points using the formula by Greville (as shown for the plane problem). For the quarter cylinder, for example, the collocation points are depicted in Figure 21.

For the implementation it is convenient to introduce a continuous numbering of the control points. We define the NURBS functions as

$$R_i = R^{p,q}_{a(i),b(i)} \tag{118}$$

and the parameters as

$$\mathbf{q}^e_i = \mathbf{q}^e_{a(i),b(i)} \tag{119}$$

$$\mathbf{d}^e_i = \mathbf{d}^e_{a(i),b(i)} \tag{120}$$

where $a(i)$ and $b(i)$ are the indices associated with point i. For the quarter cylinder example we have

$$\mathbf{a} = \begin{pmatrix} 1 & 1 & 1 & 2 & 2 & 2 \end{pmatrix} \tag{121}$$

$$\mathbf{b} = \begin{pmatrix} 1 & 2 & 3 & 1 & 2 & 3 \end{pmatrix} \tag{122}$$

With this new notation Equations 42 and 43 can be rewritten as

$$\mathbf{t} = \sum_{i=1}^{A*B} R_i(u,v)\mathbf{q}^e_i \tag{123}$$

$$\mathbf{u} = \sum_{i=1}^{A*B} R_i(u,v)\mathbf{d}^e_i \tag{124}$$

The discretised integral equation is now

$$\begin{aligned} \mathbf{c}\,(P_n) \sum_{i=1}^{A*B} R_i(s,t(P_n))\mathbf{d}_a^{ec} = & \sum_{e=1}^{E}\sum_{i=1}^{A*B}\left(\int_{S_e} \mathbf{U}\,(P_n,Q)\; R_i\; dS\right) \mathbf{q}_i^e \\ & -\sum_{e=1}^{E}\sum_{i=1}^{A*B}\left(\int_{S_e} \mathbf{T}\,(P_n,Q)\; R_i\; dS\right) \mathbf{d}_i^e \\ & \qquad for \quad n = 1,2,3...N \end{aligned} \tag{125}$$

where ec indicates the NURBS patch where the collocation point resides and E is the number of patches.

5.6 Numerical evaluation of integrals

As with the 2-D implementation we divide the integrals into regular, weakly singular and strongly singular integrals. It should be noted that the 3-D NURBS shape functions may be zero over parts of a NURBS patch, hence the parameter space u, v should be divided into a set of subregions. For the sake of simplicity this subdivision will not be discussed here.

5.7 Regular integrals

The numerical integration is carried out in the local coordinate space using Gauss Quadrature. Since Gauss formulae work with coordinates (ξ, η) that range from -1 to +1, a change from the NURBS coordinate system u, v (which has a range of 0 to 1) has to be made. Therefore two transformations are necessary. One from the global coordinates to NURBS coordinates given by

$$\mathbf{x} = \sum_{i=1}^{A*B} R_i(u,v)\mathbf{x}_{a(i),b(i)} \tag{126}$$

and one from the NURBS coordinate system u, v to ξ, η

$$\begin{aligned} u &= 0.5(\xi + 1) \\ v &= 0.5(\eta + 1) \end{aligned} \tag{127}$$

where the Jacobian is 0.25. The regular integrals over a patch e are given by

$$\triangle \mathbf{U}^e_{a,i} = \int_{-1}^{+1}\int_{-1}^{+1} \mathbf{U}(P_a, Q)\; R_i(u,v)\; J\, 0.25\; d\xi d\eta \tag{128}$$

$$= \sum_{m=1}^{M}\sum_{l=1}^{L} \mathbf{U}(P_a, Q(\xi_m, \eta_l))\; R_i(u_m, v_l)\; J\, 0.25\, W_m\, W_l \tag{129}$$

$$\triangle \mathbf{T}^e_{a,i} = \int_{-1}^{+1}\int_{-1}^{+1} \mathbf{T}(P_a, Q)\; R_i(u,v)\; J\, 0.25\; d\xi d\eta \tag{130}$$

$$= \sum_{m=1}^{M}\sum_{l=1}^{L} \mathbf{T}(P_a, Q(\xi_m, \eta_l))\; R_i(u_m, v_l)\; J\, 0.25\, W_m\, W_l \tag{131}$$

In the above M, L are the number of Gauss points in ξ, η direction (chosen depending on the proximity of the collocation point to the element), ξ_m, η_l are coordinates of Gauss points and W_m, W_l are weights. Care has to be taken when the collocation point is very near to the element as the number of Gauss points has to be increased. Sometimes it is advantageous to subdivide the integration region.

5.8 Weakly singular integrals

The kernel $\mathbf{U}$ has a singularity of $o(\frac{1}{r})$. The product $\mathbf{U}R_i$ tends to infinity if R_i does not tend to zero at the singularity point.

For this case we use a method that has been found to work well for the iso-parametric BEM, i.e. we perform the integration in a local coordinate system $(\bar{\xi}, \bar{\eta})$ where the Jacobian tends to zero as the singularity point is approached. For this we divide the integration region into triangular subregions as shown in Figure 22.

The transformations between u, v and $\bar{\xi}, \bar{\eta}$ are given by

$$u = \sum_{i=1}^{3} N_i(\bar{\xi}, \bar{\eta}) u_i \tag{132}$$

$$v = \sum_{i=1}^{3} N_i(\bar{\xi}, \bar{\eta}) v_i \tag{133}$$

where u_i, v_i are the local coordinates of the 3 nodes making the triangle, node 3 being the node where the singularity occurs. The special shape functions are

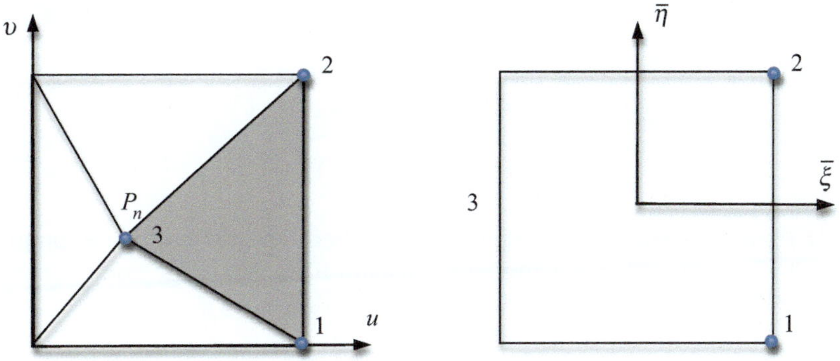

Figure 22. Coordinate transformation between u, v and $\bar{\xi}, \bar{\eta}$ systems

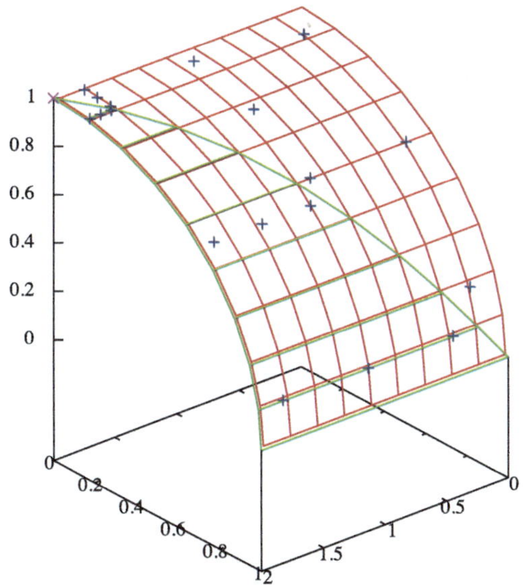

Figure 23. Location of Gauss points in global coordinate space for the case where a corner node is the collocation point

$$N_1(\bar{\xi},\bar{\eta}) = \frac{1}{4}(1+\bar{\xi})(1-\bar{\eta}) \tag{134}$$

$$N_2(\bar{\xi},\bar{\eta}) = \frac{1}{4}(1+\bar{\xi})(1+\bar{\eta}) \tag{135}$$

$$N_3(\bar{\xi},\bar{\eta}) = \frac{1}{2}(1-\bar{\xi}) \tag{136}$$

If P_n is located inside the patch, then there are 4 triangular subregions. If P_n is a corner node there are 2 regions and if P_n is an edge node, 3 regions.

The derivatives of the transformation are given by

$$\begin{aligned} \frac{\partial u}{\partial \bar{\xi}} &= \sum_{i=1}^{3} \frac{\partial N_i(\bar{\xi},\bar{\eta})}{\partial \bar{\xi}} u_i & \frac{\partial v}{\partial \bar{\xi}} &= \sum_{i=1}^{3} \frac{\partial N_i(\bar{\xi},\bar{\eta})}{\partial \bar{\xi}} v_i \\ \frac{\partial u}{\partial \bar{\eta}} &= \sum_{i=1}^{3} \frac{\partial N_i(\bar{\xi},\bar{\eta})}{\partial \bar{\eta}} u_i & \frac{\partial v}{\partial \bar{\eta}} &= \sum_{i=1}^{3} \frac{\partial N_i(\bar{\xi},\bar{\eta})}{\partial \bar{\eta}} v_i \end{aligned} \tag{137}$$

with the derivatives of the shape functions given by

$$\begin{aligned} \frac{\partial N_1(\bar{\xi},\bar{\eta})}{\partial \bar{\xi}} &= \frac{1}{4}(1-\bar{\eta}) & \frac{\partial N_1(\bar{\xi},\bar{\eta})}{\partial \bar{\eta}} &= -\frac{1}{4}(1+\bar{\xi}) \\ \frac{\partial N_2(\bar{\xi},\bar{\eta})}{\partial \bar{\xi}} &= \frac{1}{4}(1+\bar{\eta}) & \frac{\partial N_2(\bar{\xi},\bar{\eta})}{\partial \bar{\eta}} &= \frac{1}{4}(1+\bar{\xi}) \\ \frac{\partial N_3(\bar{\xi},\bar{\eta})}{\partial \bar{\xi}} &= -\frac{1}{2} & \frac{\partial N_3(\bar{\xi},\bar{\eta})}{\partial \bar{\eta}} &= 0 \end{aligned} \tag{138}$$

The Jacobian of this transformation is

$$\bar{J} = \frac{\partial u}{\partial \bar{\xi}}\frac{\partial v}{\partial \bar{\eta}} - \frac{\partial v}{\partial \bar{\xi}}\frac{\partial u}{\partial \bar{\eta}} \tag{139}$$

and tends to zero as the singular point is approached, regularizing the integral.

Applying Gauss quadrature we obtain

$$\triangle \mathbf{U}^e_{a,i} = \sum_{nt=1}^{ntr} \int_{-1}^{+1} \int_{-1}^{+1} \mathbf{U}\left(P_a, Q\right)\, R_i(u,v)\; J\, \bar{J}\; d\bar{\xi} d\bar{\eta} \tag{140}$$

$$= \sum_{nt=1}^{ntr} \sum_{m=1}^{M} \sum_{l=1}^{L} \mathbf{U}\left(P_a, Q(u_m, v_l)\right)\, R_i(u_m, v_l)\; J \bar{J}\, W_m\, W_l \tag{141}$$

where *ntr* is the number of triangular subregions.

As an example we show in Figure 23, in the global coordinate space, the location of 3x3 Gauss points for the quarter cylinder and one location of the collocation point, marked with an x. It can be seen that the density of Gauss points increases as the collocation point is approached.

Strongly singular integration

The kernel $\mathbf{T}$ has a singularity of $o(\frac{1}{r^2})$ and is therefore strongly singular.

For a *decay* infinite patch the integral to be evaluated is

$$\int_{S_e} \mathbf{T}(P_n, Q)\,(1-u)\,R_b^q\,dS \tag{142}$$

For a *plane strain* infinite patch the integral is

$$\int_{S_e} \mathbf{T}(P_n, Q)\,R_b^q\,dS \tag{143}$$

For the case where the shape function R_b^d does not tend to zero as the singularity is approached the integral of $\mathbf{T}(P_a, Q)R_i$ only exists as Cauchy principal value.

However the integrand can be made weakly singular by considering that for a rigid body motion the tractions must be zero everywhere as explained previously.

We solve the integral equation for $(\mathbf{t} = \mathbf{0})$ and 3 rigid body motions (i.e. $\mathbf{u} = \begin{pmatrix} 1 \\ 0 \\ 0 \end{pmatrix}$, $\mathbf{u} = \begin{pmatrix} 0 \\ 1 \\ 0 \end{pmatrix}$ and $\mathbf{u} = \begin{pmatrix} 0 \\ 0 \\ 1 \end{pmatrix}$) at all locations (P, Q) and compute all 9 values of the free term $\mathbf{c}$:

$$\mathbf{c}\,(P) = -\int_S \mathbf{T}\,(P, Q)\,dS \tag{144}$$

Replacing the free term in the integral equation (166) we obtain

$$-\int_S \mathbf{T}\,(P,Q)\,dS\mathbf{u}\,(P) = \int_S \mathbf{U}\,(P,Q)\,\mathbf{t}\,(Q)\,dS - \int_S \mathbf{T}\,(P,Q)\,\mathbf{u}\,(Q)\,dS \tag{145}$$

or

$$\int_S \mathbf{U}\,(P,Q)\,\mathbf{t}\,(Q)\,dS = \int_S \mathbf{T}\,(P,Q)\,(\mathbf{u}\,(Q) - \mathbf{u}(P))dS \tag{146}$$

It can be seen that as Q approaches the collocation point P, $\mathbf{T}$ is multiplied by a value approaching zero, which means the singularity is decreased to weakly singular and the integral can be computed using the method described previously.

The explanation of the rigid body process was for a finite domain and has to be extended to an infinite and semi-infinite domain.

We surround the finite surface with a spherical or hemispherical surface of radius R, where R is infinity. We have to now add in Equation (144) the integral over the spherical surface (S_p) also known as the *Azimuthal Integral* i.e.

$$\mathbf{c}(P) = -(\int_S \mathbf{T}(P,Q)\, dS + \int_{S_p} \mathbf{T}(P,Q)\, dS_p) \tag{147}$$

The *Azimuthal Integral* integral can be evaluated using polar coordinates (see Beer et al. (2008)). For an infinite domain it is given by

$$\mathbf{A} = \int_{S_p} \mathbf{T}(P,Q) dS_p = -\mathbf{I} \tag{148}$$

and for a semi-infinite domain

$$\mathbf{A} = \int_{S_p} \mathbf{T}(P,Q) dS_p = -\frac{1}{2}\mathbf{I} \tag{149}$$

where **I** is a unit matrix. Substitution into the integral equation gives

$$\int_S \mathbf{U}(P_n,Q)\,\mathbf{t}(Q)\, dS = (\int_S \mathbf{T}(P_n,Q)\, dS + \mathbf{A}))\mathbf{u}(P_n) - \int_S \mathbf{T}(P_n,Q)\,\mathbf{u}(Q)\, dS \tag{150}$$

After substitution of

$$\mathbf{u}(P_n) = \sum_{j=1}^{I_d} R_j^{d,e_n}(u_n,v_n)\mathbf{d}_j^{e_n} \tag{151}$$

(where e_n signifies the patch that contains the collocation point P_n) we obtain the discretized integral equation

$$\sum_{e=1}^{E} \triangle \mathbf{U}_n^e = (\sum_{e=1}^{E} \triangle \mathbf{T}_n^e + \mathbf{A}) \sum_{j=1}^{I_d} R_j^{d,e_n}(u_n,v_n)\mathbf{d}_j^{e_n} - \sum_{e=1}^{E}\sum_{i=1}^{I_t} \triangle \mathbf{T}_{n,i}^e \mathbf{d}_i^e \tag{152}$$

where

$$\triangle \mathbf{T}_n^e = \sum_{nt=1}^{ntr} \int_{-1}^{+1} \int_{-1}^{+1} \mathbf{T}(P_n, Q) \; J \, \bar{J} \, d\bar{\xi} d\bar{\eta} \tag{153}$$

$$= \sum_{nt=1}^{ntr} \sum_{m=1}^{M} \sum_{l=1}^{L} \mathbf{T}(P_n, Q(u_m, v_l)) \; J\bar{J} \, W_m \, W_l \tag{154}$$

and ntr is the number of triangular subregions.

5.9 Assembly

The patch contributions have to be assembled into matrices, leading to a system of equations

$$[\mathbf{T}]\{\mathbf{u}\} = [\mathbf{U}]\{\ \mathbf{t}\ \} \tag{155}$$

for finite domain problems and for infinite domain problems

$$([\mathbf{T}] + [\mathbf{I}])\{\mathbf{u}\} = [\mathbf{U}]\{\mathbf{t}\} \tag{156}$$

where in a loop over all NURBS patches E, the matrices are assembled as follows

$$\mathbf{T}_{a,inc(i)} = \mathbf{T}_{a,inc(i)} + \triangle \mathbf{T}_{a,i}^e \tag{157}$$

$$\mathbf{U}_{a,inc(i)} = \mathbf{U}_{a,inc(i)} + \triangle \mathbf{U}_{a,i}^e \tag{158}$$

where $inci(i)$ is the global number of the i-th node of patch e. The first index refers to the row and the second to the column number of the 3x3 submatrices of the matrices $[\mathbf{T}]$ and $[\mathbf{U}]$.

5.10 Solution

For a boundary value problem either $\mathbf{u}$ or $\mathbf{t}$ are known and the other unknown and can be obtained. For a tunneling problem the excavation tractions $\mathbf{t}_0$ can be determined directly from the virgin stress $\boldsymbol{\sigma}_0$. Assuming that the shear components of the virgin stress are zero, i.e.

$$\boldsymbol{\sigma}_0 = \begin{pmatrix} \sigma_{x0} \\ \sigma_{y0} \\ \sigma_{z0} \\ 0 \\ 0 \\ 0 \end{pmatrix} \tag{159}$$

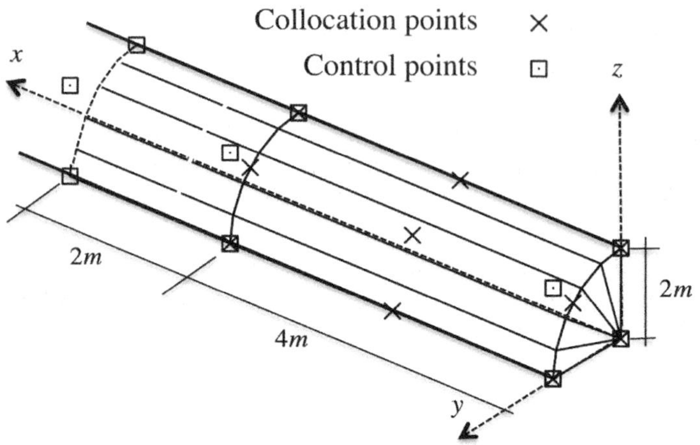

Figure 24. Discretisation of tunnel into two finite and one infinite NURBS patches for one quarter

the tractions are computed by

$$\mathbf{t}_0 = \begin{pmatrix} t_{x0} \\ t_{y0} \\ t_{z0} \end{pmatrix} = \begin{pmatrix} n_x\sigma_{x0} \\ n_y\sigma_{y0} \\ n_z\sigma_{z0} \end{pmatrix} \tag{160}$$

where n_x, n_y and n_z are components of the vector normal to the boundary.

In this case we do not need to explicitly assemble matrix $[\mathbf{U}]$ but solve the system of equations for an infinite domain

$$([\mathbf{T}] + [\mathbf{I}])\{\mathbf{u}\} = \{\mathbf{F}\} \tag{161}$$

where $\{\mathbf{F}\}$ is assembled from the following patch contributions

$$\triangle\mathbf{F}_a^e = \sum_{nt=1}^{ntr} \int_{-1}^{+1}\int_{-1}^{+1} \mathbf{U}\,(P_a, Q)\ \mathbf{t}_0\ J\ \bar{J}\ d\bar{\xi}d\bar{\eta} \tag{162}$$

$$= \sum_{nt=1}^{ntr}\sum_{m=1}^{M}\sum_{l=1}^{L} \mathbf{U}\,(P_a, Q(u_m, v_l))\ \mathbf{t}_0(u_m, v_l)\ J\bar{J}\,W_m\,W_l \tag{163}$$

where for a regular integration ntr=1 and $\bar{J}$= 1. Note that this allows for tractions which are discontinuous over the boundary.

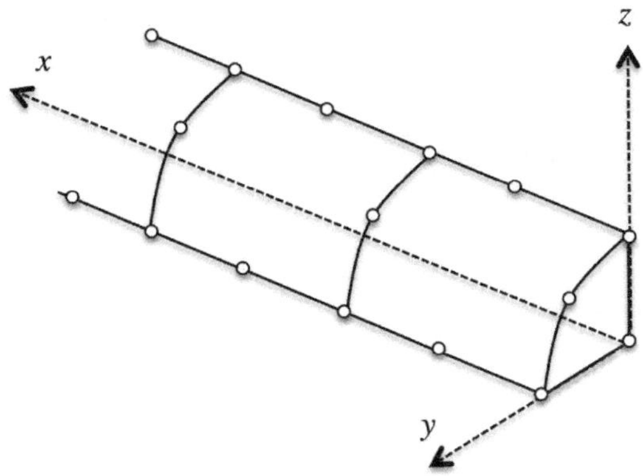

Figure 25. Discretisation of tunnel using isoparametric elements

5.11 Example

The example is an infinitely long tunnel in an infinite domain. The diameter of the tunnel is 2 m, the modulus of Elasticity 1000 MPa, the Poisson's ratio 0.3 and the virgin stress field is $\sigma_x = 0, \sigma_y = 0, \sigma_z = -1MPa$ with all shear stresses zero.

Plane strain conditions occur at two diameters away from the tunnel face and the results can be compared with the analytical solution by Kirsch (1898). The isogeometric discretization is shown in Figure 24.

Symmetry was considered using the procedure, so only 1/4 of the tunnel had to be discretized into 3 NURBS patches. Patch No. 1 is a degenerate patch modeling the face of the tunnel, patch 2 is a finite patch defining the tunnel surface and patch No 3 is an infinite *plane strain* patch. For the geometry definition the following orders were assumed:

Finite NURBS patches 1 and 2: $p = 2, q = 1$ Infinite patch 3: $q = 2$

The discretization of patch 2 exactly describes a quarter cylinder. For the variation of displacements in this patch the order for q was elevated to 2 (i.e. $q_d = 2$). The analysis comprised 30 degrees of freedom. A comparison of the radial displacements along a line 4 m from the tunnel face is made with the exact solution for a plane strain tunnel and with an isoparametric solution in Figure 26. The mesh used for the isoparametric analysis with 43 degrees of freedom is shown in Figure 25.

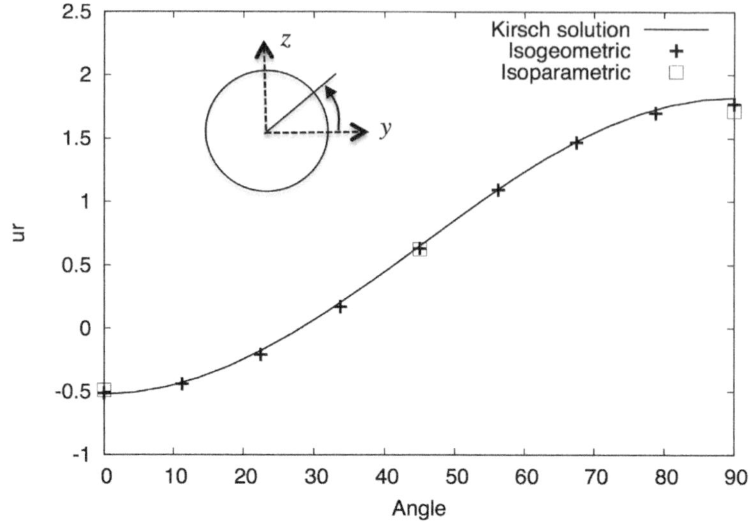

Figure 26. Comparison of radial displacement ur (mm along) a line 4 m from the tunnel face

The maximum displacement at the crown is 1.857 for the isogeometric mesh and 1.716 for the isoparametric mesh as compared with the theoretical value of 2.0. The maximum compressive stress of the isogeometric analysis is 2.996 which agrees well with the exact solution of 3.0.

5.12 Summary and Conclusions

Here we have introduced the isogeometric BEM for the solution of 3-D problems. The treatment of infinite domains as they occur in geomechanics was discussed. Special NURBS patches that are useful for such analysis (infinite and degenerate) were introduced.

6 Introduction to treatment of heterogeneous domains and non-linear material behavior

So far we have only dealt with linear elastic material behavior and homogenous domains, cases where fundamental solutions exist. The purpose of this section is to show how the BEM can be extended to deal with heterogeneous domains and non-linear material behavior, where no fundamental solutions are available. Basically, the two phenomena are associated with forces that occur inside the domain and we refer to these as body forces. Therefore, we have to augment the integral equations to include body force terms. The problem is then solved in an iterative manner.

6.1 Non-linear material behavior

We deal with non-linear material behavior in the same way as with the Finite Element Method. In the *initial stress method* the solution consists of a sequence of linear solutions, where an initial stress $\triangle\boldsymbol{\sigma}_0$ is computed and applied at each iteration. Methods of how to determine the initial stresses arising from various types of elasto-plastic behavior will not be discussed here. More details can be found in Ribeiro et al. (2008) . We use the Voight notation for stress and strain here and this means that for 2-D problems the stress pseudo-vector is defined as

$$\triangle\boldsymbol{\sigma}_0 = \begin{pmatrix} \sigma_x \\ \sigma_y \\ \tau_{xy} \end{pmatrix} \tag{164}$$

6.2 Heterogeneous domains

Heterogeneous domains can be treated in a similar way as nonlinear material behavior by a sequence of linear solutions where initial stresses are applied. For a piecewise heterogenous domain we can visualize a homogeneous domain having a number of inclusions. For each inclusion we can compute an initial stress as (see Riederer et al. (2009))

$$\triangle\boldsymbol{\sigma}_0 = (\mathbf{D}_{incl} - \mathbf{D})\triangle\boldsymbol{\epsilon} \tag{165}$$

where $\mathbf{D}$ is the constitutive matrix for the homogeneous domain (for which the fundamental solution has been calculated) and $\mathbf{D}_{incl}$ is the one for the inclusion. $\triangle\boldsymbol{\epsilon}$ is the increment of strain.

6.3 Boundary integral equation with body forces

The boundary integral equation for an elastic continuum with body forces can be written as (see also Beer et al. (2008)):

$$\mathbf{cu} = \int_S \mathbf{U}\,\mathbf{t}dS - \int_S \mathbf{T}\,\mathbf{u}dS + \int_V \mathbf{E}\triangle\boldsymbol{\sigma}_0 dV \tag{166}$$

where the newly introduced matrix $\mathbf{E}$ contains the derived fundamental solution for the strain. For a 2-D problem we have

$$\mathbf{E} = \begin{pmatrix} \frac{dU_{xx}}{dx} & \frac{dU_{xy}}{dx} \\ \frac{dU_{yx}}{dx} & \frac{dU_{yy}}{dy} \\ \frac{dU_{xx}}{dy} + \frac{dU_{yx}}{dx} & \frac{dU_{xy}}{dy} + \frac{dU_{yy}}{dx} \end{pmatrix} \tag{167}$$

and $\triangle\boldsymbol{\sigma}_0$ are the initial stresses.

Using the *collocation method* we obtain a system of equations of the type

$$\mathbf{cu}(P_a) = \int_S \mathbf{U}(P_a,Q)\,\mathbf{t}dS - \int_S \mathbf{T}(P_a,Q)\,\mathbf{u}dS + \int_V \mathbf{E}(P_a,\bar{Q})\triangle\boldsymbol{\sigma}_0 dV \tag{168}$$

6.4 Discretisation

In addition to the surface discretisation already discussed we require a volume discretisation for the evaluation of the volume integrals. There are two approaches that can be taken:

- Cells that are body fitted to the boundary
- Cartesian call grid

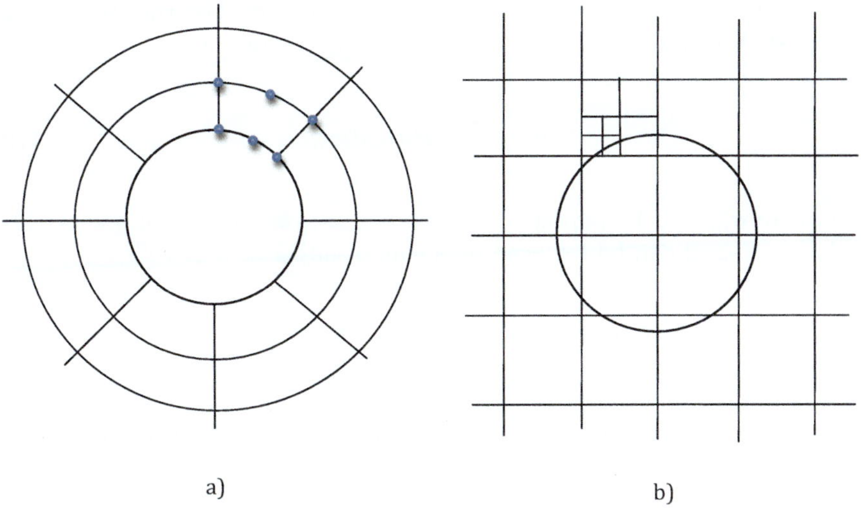

Figure 27. a) Bodyfitted cells and b) Cartesian grid

The two approaches are compared in Figure 27

In the first approach we generate a cell mesh that is fitted to the boundary discretisation. In the second approach we establish a regular Cartesian cell mesh, which is independent of the boundary discretisation. The advantage of the second approach is that no mesh generation is needed.

Although the cell mesh looks very much like a finite element mesh there are distinct differences. Firstly, the mesh is only used to evaluate the volume integral and is not associated with any degrees of freedom. Secondly, the cell mesh need only be extended to the area where initial stresses occur (i.e. only in areas where elasto-plastic behavior occurs or where inclusions are located).

Body fitted cells The idea here is that cell boundaries are fitted to the elements that describe the boundary of the problem. For 2-D problems we may use the tensor product of NURBS (as introduced in lecture 4) for the description of the geometry of the cell.

$$\mathbf{x} = \sum_{i=1}^{A*B} R_i(u,v)\mathbf{x}_i^c \tag{169}$$

where $\mathbf{x}_i^c$ are control point coordinates of cell c. For the variation of initial stresses we may assume the same interpolation, i.e.

$$\boldsymbol{\sigma}_0 = \sum_{i=1}^{A*B} R_i(u,v)\boldsymbol{\sigma}_{0i}^c \tag{170}$$

where $\boldsymbol{\sigma}_{0i}^c$ are the parameters for initial stresses at control points. The volume integral can be replaced by a sum of integrals over cells (c)

$$\int_V \mathbf{E}(P_a,\bar{Q})\triangle\boldsymbol{\sigma}_0 dV = \sum_{c=1}^{C}\sum_{i=1}^{A*B} \triangle\mathbf{E}_{a,i}^e \boldsymbol{\sigma}_{0i}^c \tag{171}$$

where the integration over cells is carried out numerically using Gauss Quadrature.

$$\begin{aligned} \triangle\mathbf{E}_{a,i}^c &= \int_{-1}^{+1}\int_{-1}^{+1} \mathbf{E}\left(P_a,Q\right)\; R_i\; J\, 0.25\; d\xi d\eta & (172)\\ &= \sum_{m=1}^{M}\sum_{l=1}^{L} \mathbf{E}\left(P_a,Q(\xi_m,\eta_l)\right)\; R_i\; J\, 0.25\; W_m\; W_l & (173)\end{aligned}$$

The problem with this approach is that in oder to compute $\boldsymbol{\sigma}_{0i}^c$, boundary stresses must be computed at each iteration. Because of the hypersingularity of the Kernel this is difficult. In addition this approach would involve an additional mesh generation effort.

Cartesian grid In this approach we define a regular grid that only exists in computer memory. The grid will contain cells that either intersect the boundary or are outside the domain. These cells have to be detected. Using the outward normal vector a simple algorithm can be developed to detect cells that are outside the domain. These are eliminated. For cells that cut the boundary we develop a method where cells are subdivided using an Octree method (shown in Figure 1 for one of the cells). Cells where the major part is outside the domain are eliminated. The level of refinement near the boundary and therefore accuracy of the method can be controlled by the user. In the simplest implementation we may assume a constant distribution of the initial stress inside a cell. The advantage of this is that because points are not on the boundary, the computation of internal stresses does not involve a hypersingular Kernel.

For constant cells the volume integral can be replaced by:

$$\int_V \mathbf{E}(P_a, Q)\triangle\boldsymbol{\sigma}_0 dV = \sum_{c=1}^{C} \triangle\mathbf{E}_a^c\boldsymbol{\sigma}_0^c \tag{174}$$

where the sum is over all valid cells C, $\boldsymbol{\sigma}_0^c$ is the initial stress in the cell c and

$$\triangle\mathbf{E}_a^c = \int_{-1}^{+1}\int_{-1}^{+1} \mathbf{E}(P_a, Q)\; J\,0.25\, d\xi d\eta \tag{175}$$

$$= \sum_{m=1}^{M}\sum_{l=1}^{L} \mathbf{E}(P_a, Q(\xi_m, \eta_l))\; J\,0.25\;W_m\,W_l \tag{176}$$

6.5 Solution Process

The assembled system of equations with the body force term is given by

$$[\mathbf{T}]\{\mathbf{u}\} = [\mathbf{U}]\{\mathbf{t}\} + \{\mathbf{F}\}_0 \tag{177}$$

where

$$\{\mathbf{F}\}_0 = \begin{pmatrix} \sum_{c=1}^{C} \triangle\mathbf{E}_1^c\boldsymbol{\sigma}_0^c \\ \sum_{c=1}^{C} \triangle\mathbf{E}_2^c\boldsymbol{\sigma}_0^c \\ \vdots \end{pmatrix} \tag{178}$$

The solution is iterative, starting with an elastic solution without body forces (i.e. $\{\mathbf{F}\}_0 = 0$). After this solution the initial stresses are computed at the cells. A new right hand side ($\{\mathbf{F}\}_0$) is assembled and a new increment of displacement is computed. The iteration can be stopped if a suitable norm of ($\{\mathbf{F}\}_0$) is smaller than a specified value. In fact, the solution procedure is identical to that used in the FEM.

6.6 Example

This example is taken from Riederer (2010) and is related to a tunnel with rock bolts and soft inclusion zone. Both are treated as inclusions. The tunnel radius is $r = 5m$, the overburden is $h = 795m$, the soft zone is $2m$ thick and the distance to the tunnel is $1m$. The Rock mass properties are: $E = 10000MN/m^2$, $\nu = 0.333$, $\gamma = 0.02MN/m^3$. The Inclusion properties are: $E = 10MN/m^2$, $\nu = 0.333$, $\gamma = 0.02MN/m^3$. The Bolt properties are: $E = 200000MN/m^2$, $A = 0.001963m^2$.

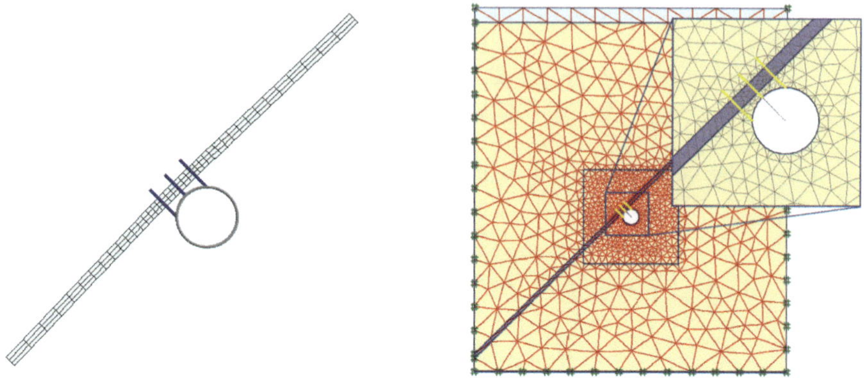

Figure 28. Left: BEM mesh and right FEM mesh for tunnel problem

The results are compared with those calculated by the FEM-program PLAXIS. The FEM and BEM discretisations are presented in Figure 28 it can be seen that the FE-mesh is much more complex than the BE-mesh. The deformed mesh and the contours of the total displacements are shown in Figure 29. A comparison of the results (displacements along rock bolts) is made in Figure 30. It can be seen that the results of both methods agree well.

7 Summary and conclusions

In this chapter an introduction was given to the implementation of the isogeometric Boundary Element Method. The method is an ideal companion to CAD, as both define geometries by its surfaces. This opens the possibility for completely avoiding mesh generation as geometry information can be taken directly from CAD data. The isogeometric implementation of the BEM has other benefits too. These include an improved quality of the results because of the higher continuity of the basis functions and efficient refinement strategies. Regarding the refinement, a new strategy was introduced that keeps the approximation of the geometry (that is exactly as in the CAD representation) unchanged and only improves the approximation of the unknown. The examples presented show that with the same number of unknowns, results of the isogeometric BEM are superior to the results of the classical isobarmetric BEM.

At the end of the chapter we have shown that the method can be ex-

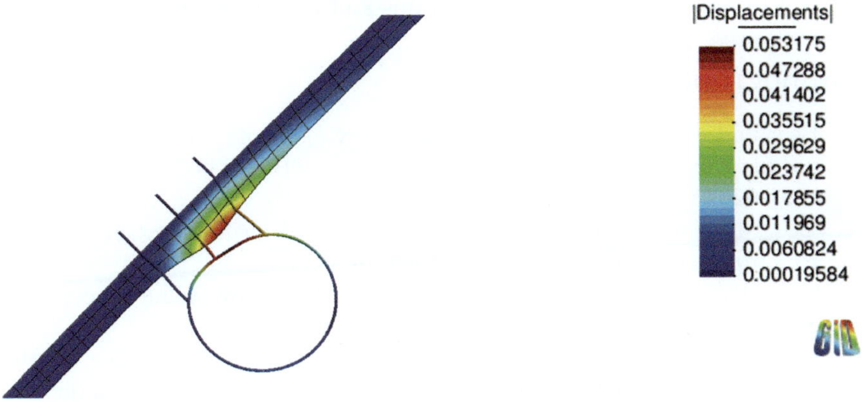

Figure 29. Deformed mesh with contours of resultant displacements

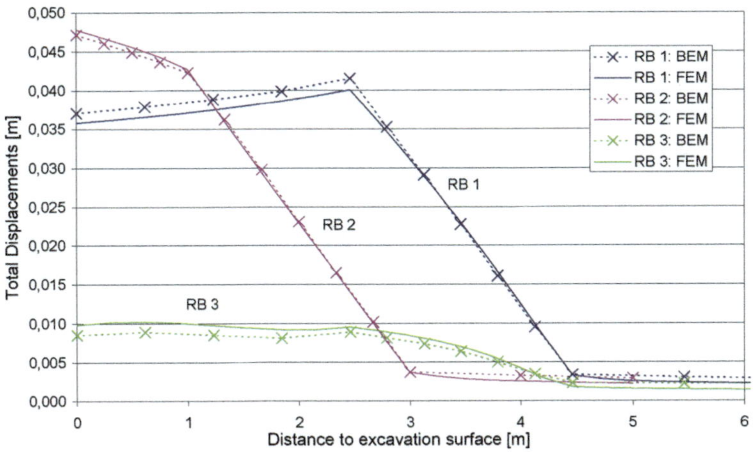

Figure 30. Comparison of results: Displacements along rock bolts (RB) 1-3

tended to handle heterogeneous domains and non-linear material behavior.

Bibliography

G. Beer and J. O. Watson. Infinite boundary elements. *International Journal for Numerical Methods in Engineering*, 28(6):1233–1247, 1989. ISSN 1097-0207.

G. Beer, I. Smith, and C. Duenser. *The Boundary Element Method with Programming*. Springer-Verlag, Wien, 2008.

T. Greville. Numerical procedures for interpolation by spline functions. *J. Soc. Ind. Appl. Math. Ser. B. Numer. Anal.*, 1964.

Peter R. Johnston and David Elliott. A generalisation of telles' method for evaluating weakly singular boundary element integrals. *Journal of Computational and Applied Mathematics*, 131(1–2):223–241, 2001.

Ernst Gustav Kirsch. Die Theorie der Elastizität und die Bedürfnisse der Festigkeitslehre. *Zeitschrift des Vereines deutscher Ingenieure*, 42:797–807, 1898.

Les Piegl and Wayne Tiller. *The NURBS book (2nd ed.)*. Springer-Verlag New York, Inc., New York, NY, USA, 1997. ISBN 3-540-61545-8.

T. Ribeiro, G. Beer, and C. Duenser. Efficient elastoplastic analysis with the boundary element method. *Computational Mechanics*, 41:715–732, 2008. ISSN 0178-7675.

K. Riederer. *Modelling of Ground Support in Tunnelling using the BEM*. PhD thesis, Graz University of Technology, 2010.

K. Riederer, C. Duenser, and G. Beer. Simulation of linear inclusions with the BEM. *Engineering Analysis with Boundary Elements*, 33(7):959 – 965, 2009. ISSN 0955-7997.

Ch. Duenser W. Moser and G. Beer. Mapped infinite elements for three-dimensional multi-region boundary element analysis. *International Journal for Numerical Methods in Engineering*, 61:317–328, 2004.

An Introduction to Isogeometric Collocation Methods

Alessandro Reali[1,2*] and Thomas J.R. Hughes[3†]

[1] Department of Civil Engineering and Architecture, University of Pavia, Italy
[2] CNR Institute of Applied Mathematics and Information Technology "E. Magenes", Pavia, Italy
[3] Institute for Computational Engineering and Sciences, University of Texas at Austin, USA.

Abstract Within the framework of isogeometric analysis, collocation methods have been recently proposed as an interesting strong form alternative to standard Galerkin approaches, characterized by a significantly reduced computational cost, but still guaranteeing higher order convergence rates. In this chapter we provide a concise introduction to isogeometric collocation methods and propose a brief review of some of the most important results obtained so far in this context.

1 Introduction

The main aim of Isogeometric analysis (IGA), introduced ten years ago by Hughes et al. (2005), is to bridge the gap between Computer Aided Design (CAD) and Finite Element Analysis (FEA). Accordingly, the basic IGA paradigm consists of adopting the same basis functions used for geometry representations in CAD systems - such as, e.g., Non-Uniform Rational B-Splines (NURBS) - for the approximation of field variables, in an isoparametric fashion. This leads to a cost-saving simplification of the typically expensive mesh generation and refinement processes required by standard FEA. Moreover, thanks to the high-regularity properties of its basis functions, IGA has shown a better accuracy per-degree-of-freedom and an enhanced robustness with respect to standard FEA in a number of

*A. Reali was supported by the European Research Council (ERC grant "ISOBIO", agreement no. 259229).

†T.J.R. Hughes was supported by grants from the Office of Naval Research (N00014-08-1-0992), the National Science Foundation (CMMI-01101007), and SINTEF (UTA10-000374) with the University of Texas at Austin.

G. Beer, S. Bordas (Eds.), *Isogeometric Methods for Numerical Simulation*,
CISM International Centre for Mechanical Sciences
DOI 10.1007/ 978-3-7091-1843-6_4 © CISM Udine 2015

applications ranging from solids and structures (see, e.g., Auricchio et al., 2010b; Borden et al., 2012; Caseiro et al., 2014; Cottrell et al., 2006, 2007; de Falco et al., 2011; Dhote et al., 2014; Elguedj et al., 2008; Hughes et al., 2008, 2014; Lipton et al., 2010; Morganti et al., 2014; Reali, 2006) to fluids (see, e.g., Akkerman et al., 2007; Bazilevs et al., 2007; Bazilevs and Hughes, 2008; Buffa et al., 2011; Liu et al., 2014; Gomez et al., 2010), opening also the door to geometrically flexible discretizations of higher-order partial differential equations in primal form (see, e.g., Auricchio et al., 2007; Gomez et al., 2008; Kiendl et al., 2009).

It is well known that an important issue related to IGA concerns the development of efficient integration rules, able to reduce assembly costs in particular when higher-order approximations are employed. The fact that element-wise Gauss quadrature, typically used for standard FEA and originally adopted for Galerkin-based IGA, does not properly take into account the inter-element higher continuity of the IGA basis functions leads to suboptimal assembly costs, significantly affecting the performance of IGA methods. Ad hoc quadrature rules have been proposed by Hughes et al. (2010), Auricchio et al. (2012b), and Schillinger et al. (2014a), but the development of a general and effective solution for Galerkin-based IGA methods is still an open problem.

Aiming at optimizing the computational cost, still taking advantage of IGA geometrical flexibility and accuracy, isogeometric collocation schemes have been recently proposed by Auricchio et al. (2010a, 2012a). The fundamental idea consists of the discretization of the governing partial differential equations in strong form, adopting the isoparametric paradigm and making use of the higher-continuity properties of the IGA shape functions. Detailed comparisons with both IGA and FEA Galerkin-based approaches have shown IGA collocation advantages in terms of accuracy versus computational cost, in particular when higher-order approximation degrees are adopted (Schillinger et al., 2013). In general, IGA collocation features look particularly desirable in all those situations where evaluation and assembly costs are dominant, as in the case of explicit structural dynamics where the computational cost is dominated by stress divergence evaluations at quadrature points for the calculation of the residual force vector (Auricchio et al., 2012a; Schillinger et al., 2013).

Within the IGA collocation context, several promising significant studies have been recently published, including phase-field modeling (Gomez et al., 2014), contact (De Lorenzis et al., 2014b), and hierarchical local refinement (Schillinger et al., 2013). Moreover, IGA collocation has been very successful in the context of structural elements. In particular, Bernoulli-Euler beam and Kirchhoff plate elements have been proposed by Reali and Gomez

(2014), and shear-deformable structural elements have been considered in a number of papers. Mixed formulations both for Timoshenko initially-straight planar beams (Beirão da Veiga et al., 2012) and for curved spatial rods (Auricchio et al., 2013) have been proposed and studied. Isogeometric collocation has been moreover successfully applied to the solution of Reissner-Mindlin plate problems in Kiendl et al. (2014a). Finally, a new single-parameter formulation for shear-deformable beams, recently introduced by Kiendl et al. (2014b), has been solved also via IGA collocation.

It is therefore clear that, thanks to its interesting and promising features, IGA collocation has attracted and is increasingly attracting a good deal of interest among Computational Mechanics researchers. This chapter is aimed to people who desire to get acquainted with the basic concepts of IGA collocation, and its goal is to provide a concise introduction, as well as to propose a brief review of some of the most important results obtained so far in this context.

2 Basics of NURBS-Based IGA Collocation

In this section, we aim at introducing the basic concepts of NURBS-based IGA collocation. After some preliminaries on B-Splines and NURBS, we introduce the idea of collocation in a very simple 1D setting, presenting also some theoretical and numerical results.

2.1 B-Spline and NURBS preliminaries

In the following, we briefly introduce the basic definitions and notations about B-Splines and NURBS. For further details, readers are referred to Piegl and Tiller (1997), Cottrell et al. (2009), and references therein.

A B-Spline basis function of degree p is generated starting from a non-decreasing sequence of real numbers referred to as knot vector

$$\Xi = \{\xi_1, ..., \xi_{m+p+1}\} \tag{1}$$

where m is the number of basis functions (equal to the number of the associated control points). A univariate B-Spline basis function $N_{i,p}(\xi)$ can be then constructed using the following Cox-de Boor recursion formula: Starting from $p = 0$, where

$$N_{i,0}(\xi) = \begin{cases} 1 & \xi_i \leq \xi < \xi_{i+1} \\ 0 & \text{otherwise} \end{cases} \tag{2}$$

the basis functions for $p > 0$ are obtained from

$$N_{i,p}(\xi) = \frac{\xi - \xi_i}{\xi_{i+p} - \xi_i} N_{i,p-1}(\xi) + \frac{\xi_{i+p+1} - \xi}{\xi_{i+p+1} - \xi_{i+1}} N_{i+1,p-1}(\xi) \tag{3}$$

where the convention $0/0 = 0$ is assumed. Given the multiplicity k of a knot, the smoothness of the B-Spline basis is C^{p-k} at that location, while it is C^{∞} everywhere else. In so-called open knot vectors, the first and the last knots have multiplicity $k = p + 1$ and the basis is interpolatory at the ends (see, e.g., Figure 1).

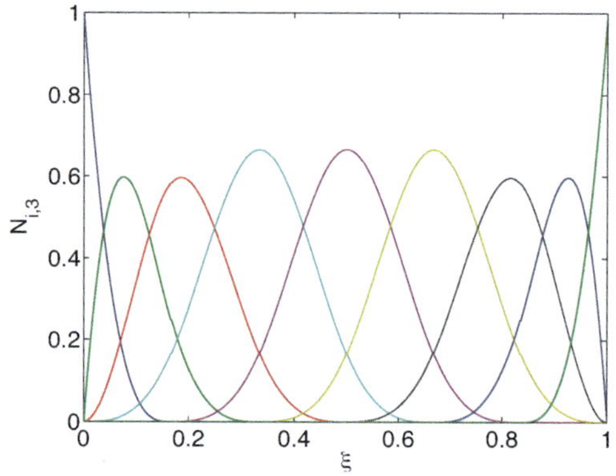

Figure 1. Cubic B-Spline basis functions formed from the open knot vector $\Xi = \{0, 0, 0, 0, 1/6, 1/3, 1/2, 2/3, 5/6, 1, 1, 1, 1\}$.

A B-Spline curve can be then constructed as the linear combination of the basis functions

$$\mathbf{C}(\xi) = \sum_{i=1}^{m} N_{i,p}\mathbf{P}_i \tag{4}$$

where the coefficients $\mathbf{P}_i \in \mathbb{R}^{d_s}$ of the linear combination are the so-called control points, being d_s the dimension of the physical space (see, e.g., Figure 2).

Multivariate B-Splines are generated through the tensor product of univariate B-Splines. If d_p denotes the dimension of the parametric space, d_p univariate knot vectors are needed:

$$\Xi^d = \left\{\xi_1^d, ..., \xi_{m_d+p_d+1}^d\right\} \tag{5}$$

where $d = 1, ..., d_p$, p_d is the polynomial degree in the parametric direction d, and m_d is the associated number of basis functions. Denoting the univariate

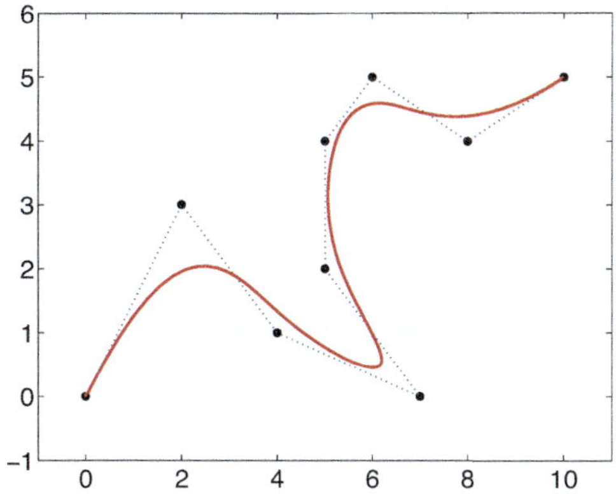

Figure 2. A 2D piecewise cubic B-Spline curve generated from the basis functions of Figure 1 (solid line), along with its control points (black dots), and control net (dotted line)

basis functions in each parametric direction ξ^d by $N^d_{i_d,p_d}$, the multivariate basis functions $B_{\mathbf{i},\mathbf{p}}(\boldsymbol{\xi})$ are obtained as

$$B_{\mathbf{i},\mathbf{p}}(\boldsymbol{\xi}) = \prod_{d=1}^{d_p} N^d_{i_d,p_d}\left(\xi^d\right) \tag{6}$$

where the multi-index $\mathbf{i} = \left\{i_1, ..., i_{d_p}\right\}$ denotes the position in the tensor product structure, $\mathbf{p} = \{p_1, ..., p_d\}$ indicates the polynomial degrees, and $\boldsymbol{\xi} = \left\{\xi^1, ..., \xi^{d_p}\right\}$ is the vector of the parametric coordinates in each parametric direction d. B-Spline surfaces and solids are obtained, for $d_p = 2$ and $d_p = 3$, respectively, from a linear combination of multivariate B-Spline basis functions and control points as follows

$$\mathbf{S}(\boldsymbol{\xi}) = \sum_{\mathbf{i}} B_{\mathbf{i},\mathbf{p}}(\boldsymbol{\xi})\,\mathbf{P}_{\mathbf{i}} \tag{7}$$

where the summation is extended to all combinations of the multi-index $\mathbf{i}$.

NURBS basis functions in $\mathbb{R}^{d_s}$ are obtained from a projective transformation of their B-Spline counterparts in $\mathbb{R}^{d_s+1}$. Univariate NURBS basis functions $R_{i,p}(\xi)$ are given by

$$R_{i,p}(\xi) = \frac{N_{i,p}(\xi)\,w_i}{\sum_{j=1}^{m} N_{j,p}(\xi)\,w_j} \tag{8}$$

where $N_{i,p}$ are B-Spline basis functions and w_i are the corresponding weights (i.e., the (d_s+1)-th coordinates of the B-Spline control points in $\mathbb{R}^{d_s+1}$). Finally, multivariate NURBS basis functions are obtained as

$$R_{\mathbf{i},\mathbf{p}}(\boldsymbol{\xi}) = \frac{B_{\mathbf{i},\mathbf{p}}(\boldsymbol{\xi})\, w_{\mathbf{i}}}{\sum_{\mathbf{j}} B_{\mathbf{j},\mathbf{p}}(\boldsymbol{\xi})\, w_{\mathbf{j}}} \tag{9}$$

and NURBS surfaces and solids are constructed as

$$\mathbf{S}(\boldsymbol{\xi}) = \sum_{\mathbf{i}} R_{\mathbf{i},\mathbf{p}}(\boldsymbol{\xi})\, \mathbf{P}_{\mathbf{i}} \tag{10}$$

2.2 IGA collocation in 1D

In this section, we introduce the basic ideas of IGA collocation in a very simple 1D setting. We also discuss the choice of collocation points and propose some theoretical results. We then conclude the section presenting some numerical tests.

Formulation. Let f, a_0, a_1, be real functions in $C^0[a,b]$, with $a<b$ given real numbers. Let $g_0, g_1 \in \mathbb{R}$ be scalars and $\mathcal{BC}_0, \mathcal{BC}_1 : C^1[a,b] \to \mathbb{R}$ be linear operators. We are interested in the following simple one-dimensional model differential problem. Find a real function $u \in C^2[a,b]$ such that

$$\begin{cases} u''(x) + a_1(x)u'(x) + a_0(x)u(x) = f(x) & \forall x \in (a,b) \\ \mathcal{BC}_i(u) = g_i & i = 0,1 \end{cases} \tag{11}$$

where u', u'' represent the first and second derivatives of u, respectively (we note that in the following we will indicate the derivative operator of order i also as D^i, $i \in \mathbb{N}$). We assume that (11) has one and only one solution u, and that the boundary condition operators $\mathcal{BC}_i$ are linearly independent on $\mathrm{Ker}(D^2)$, that is, on the space of linear functions.

To discretize problem (11) via IGA collocation, we proceed as follows. Given $n \in \mathbb{N}$, let $\mathcal{V}_{n+2} \subset C^2[a,b]$ be a NURBS space of dimension $n+2$ on the interval $[a,b]$, associated with a spline space $\hat{\mathcal{S}}_{n+2} \subset C^2[0,1]$ on the parametric interval $[0,1]$. With standard assumptions on the one-dimensional geometrical map F, we consider $DF > 0$ on the parametric domain $[0,1]$. Given, for all $n \in \mathbb{N}$, $\tau_1 < \tau_2 < ... < \tau_n$ assigned collocation points in $[a,b]$, we obtain the following discrete problem: Find $u_n \in \mathcal{V}_{n+2}$ such that

$$\begin{cases} u_n''(\tau_j) + a_1(\tau_j)u_n'(\tau_j) + a_0(\tau_j)u_n(\tau_j) = f(\tau_j) & j = 1,..,n \\ \mathcal{BC}_i u_n = g_i \qquad i = 0,1 \end{cases} \tag{12}$$

It is to be remarked that this formulation gives rise in general to a non-symmetric (but diagonally dominant) system matrix.

Collocation points and theoretical results. The discrete problem (12) is defined once a strategy for the selection of the n collocation points is set. Such a selection is of paramount importance, because it directly influences the stability and convergence properties of the collocation scheme.

In the IGA collocation literature, the images of so-called Greville abscissae (see de Boor, 2001) have been widely adopted as the default choice for collocation points. Greville abscissae are n points easily defined from the knot vector as

$$\overline{\xi}_i = (\xi_{i+1} + \xi_{i+2} + \ldots + \xi_{i+p})/p \tag{13}$$

and are well known in the CAD literature for a number of properties, among which the fact that they typically give a stable interpolation (except in some cases when high degrees are combined with particular non-uniform meshes[1]).

The selection of points guaranteeing a stable interpolation is a fundamental issue for a collocation scheme, since it is proven in Auricchio et al. (2010a) that this implies optimal convergence (i.e., of order $p-1$) in the $W^{2,\infty}$-norm (or, equivalently, in the H^2-norm).

Unfortunately, such a proof is valid only in 1D and cannot be extended to higher dimensions. However, as we will also see in the following sections, extensive numerical testing has shown that the convergence rates obtained in 1D are attained also in higher dimensions.

Moreover, optimal convergence rates are not recovered in the L^∞- and $W^{1,\infty}$-norms (or, equivalently, in the L^2- and H^1-norms), where it has been numerically shown (see, e.g., next section) that orders of convergence p and $p-1$ for even and odd degrees, respectively, are attained.

It is important to note that, despite not being optimal in the L^2- and H^1-norms (except in the case of the H^1-norm for p even) as it happens instead for Galerkin methods, the obtained orders of convergence are increasing with p, whereas the cost of collocation is much lower than that of Galerkin approaches of the same order, especially as p increases. This makes IGA collocation very competitive with respect to Galerkin on the basis of an accuracy-to-computational-cost ratio, in particular when higher degrees (e.g., $p > 3$) are adopted. More details in this sense will be given later in the framework of 3D elasticity, and interested readers are referred to the comprehensive study reported in Schillinger et al. (2013).

[1] An alternative that is proven to be always stable is given by the so-called Demko abscissae (Demko, 1985), which can be computed by an iterative algorithm (see Auricchio et al., 2010a, for more details and for a discussion on their use within IGA collocation).

Numerical results. We now present some numerical experiments in 1D confirming the convergence rates previously discussed.

In particular, we consider the following model problem defined on the domain $[0, 1]$:

$$\begin{cases} -u'' + u' + u = (1 + 4\pi^2)\sin(2\pi x) - 2\pi\cos(2\pi x), & \forall x \in (0, 1) \\ u(0) = u(1) = 0 \end{cases} \tag{14}$$

which admits the exact solution:

$$u = \sin(2\pi x) \tag{15}$$

This problem is numerically solved using the collocation method outlined in the previous sections, adopting Greville abscissae as collocation points. In Figures 3-5, we report log-scale plots of the relative errors for different degrees of approximations in L^∞-, $W^{1,\infty}$-, and $W^{2,\infty}$-norms, respectively. The obtained results show that in the first two norms an order of convergence p is attained for even degrees, while an order $p-1$ is attained for odd degrees. Instead in the $W^{2,\infty}$-norm, we observe the expected optimal order of convergence, i.e., $p - 1$, for all approximation degrees, in agreement with the predictions of the theory[2].

More results, considering the influence of a nonlinear parameterization, of different choices of collocation points like Demko abscissae, and of Neumann boundary conditions, may be found in Auricchio et al. (2010a).

Spectral approximation. One of the most important results of Galerkin IGA, with a fundamental impact on the solution of structural dynamics problems, is its capability of approximating higher modes, without introducing spurious "optical branches" in the numerical spectrum (see Cottrell et al., 2006; Reali, 2006; Hughes et al., 2008, 2014). The same capabilities are present also in IGA collocation, as it has been shown in Auricchio et al. (2010a). To this end, the following 1D eigenvalue problem is considered

$$\begin{cases} u'' + \omega^2 u = 0 & \forall x \in (0, 1) \\ u(0) = u(1) = 0 \end{cases} \tag{16}$$

[2] In Figures 3-5, we have not included results for $p = 2$, since this case is not covered by the theory of Auricchio et al. (2010a). However, the results obtained are in complete agreement with the ones observed for higher degrees, i.e., order p for the L^∞- and $W^{1,\infty}$-norms and $p - 1$ for the $W^{2,\infty}$-norm (see also the numerical results of next section).

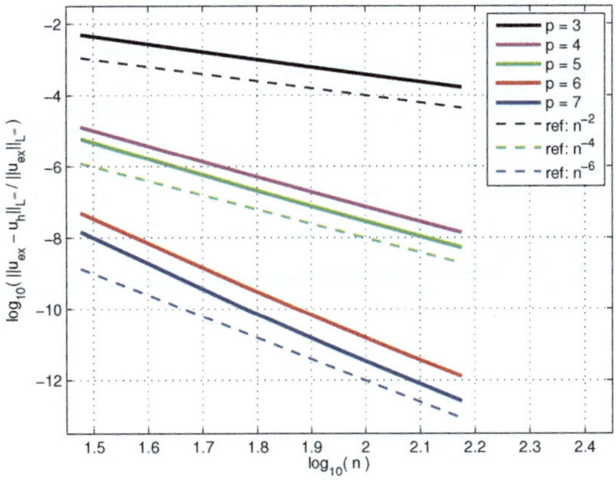

Figure 3. 1D model problem. Relative error in L^∞-norm for different degrees of approximation.

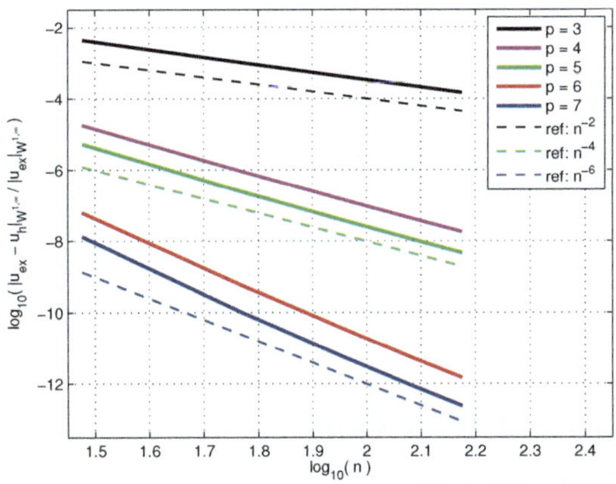

Figure 4. 1D model problem. Relative error in $W^{1,\infty}$-norm for different degrees of approximation.

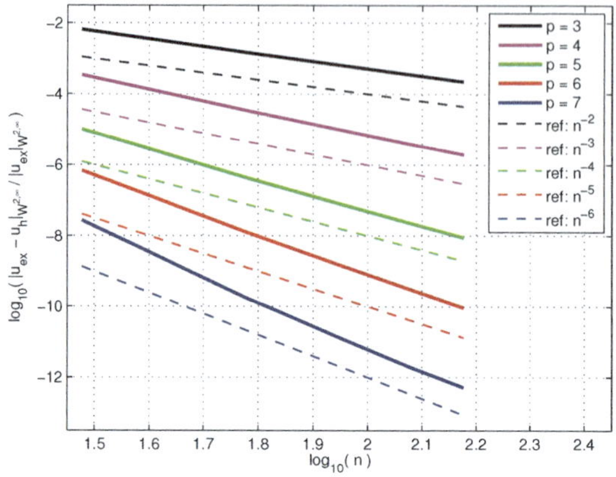

Figure 5. 1D model problem. Relative error in $W^{2,\infty}$-norm for different degrees of approximation.

for which the exact frequencies ω_n are given by

$$\omega_n = 2\pi n \qquad \text{with } n = 1, 2, 3, \ldots \tag{17}$$

Problem (16) is solved using the collocation method with Greville abscissae and, in Figure 6, we report the results in terms of normalized discrete spectra, obtained considering a linear parameterization and using different degrees of approximation (1000 d.o.f.'s have been used to produce each spectrum). It is possible to observe the good behavior of all spectra, which converge for an increasing degree p as it happens with Galerkin IGA (for more details on eigenvalue problems solved via IGA collocation, see Auricchio et al., 2010a).

3 NURBS-Based IGA Collocation for Linear Elastostatics

In this section, we extend the previously introduced IGA collocation methods to the multi-dimensional case, considering in particular linear elastostatics as model problem. Accordingly, in the following, we present the basic equations of linear elastostatics and introduce an IGA collocation formulation to solve them. Several numerical tests are proposed to show the behavior of the proposed formulation, as well as to compare its perfor-

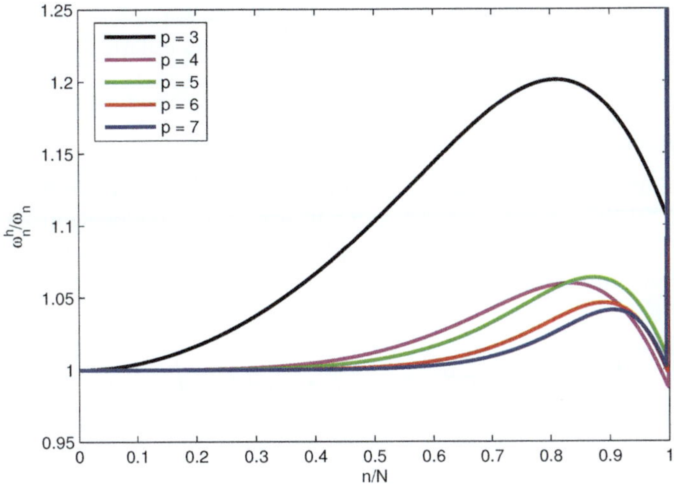

Figure 6. 1D eigenvalue problem with linear parameterization using Greville abscissae. Normalized spectra for different degrees of approximation.

mance with that of Galerkin IGA and FEA on the basis of the accuracy-to-computational-cost ratio.

3.1 Linear elastostatics

Let $\Omega \subset \mathbb{R}^{d_s}$ represent an elastic body $\mathcal{B}$ subjected to body forces $\mathbf{f}$, to prescribed displacements $\mathbf{g}$ on a portion of the boundary Γ_g, and to (possibly zero) prescribed tractions $\mathbf{p}$ on the remaining portion Γ_p, being $\Gamma = \Gamma_g \bigcup \Gamma_p$ the boundary of the domain, with $\Gamma_g \bigcap \Gamma_p = \emptyset$. Suitable regularity requirements are assumed to hold for $\mathbf{f}$, $\mathbf{g}$, and $\mathbf{p}$.

The small-strain linear elastostatics problem in strong form is defined as

$$\nabla \cdot \left(\mathbb{C}\nabla^S \mathbf{u}\right) + \mathbf{f} = \mathbf{0} \qquad \text{in}\, \Omega \tag{18}$$

complemented by the Dirichlet boundary conditions

$$\mathbf{u} = \mathbf{g} \qquad \text{on}\, \Gamma_g \tag{19}$$

and by the Neumann boundary conditions

$$\left(\mathbb{C}\nabla^S \mathbf{u}\right) \cdot \mathbf{n} = \mathbf{p} \qquad \text{on}\, \Gamma_p \tag{20}$$

where, $\mathbf{u}(\mathbf{x})$ is the unknown displacement field ($\mathbf{x}$ being the position vector), ∇ is the standard nabla operator and ∇^S is its symmetric part, $\mathbb{C}$ is the fourth-order elasticity tensor, and $\mathbf{n}$ is the unit outward normal to the boundary of the domain.

3.2 IGA collocation for elastostatics

As in Auricchio et al. (2012a), Schillinger et al. (2013), and De Lorenzis et al. (2014b), herein, we choose to interpret the collocation method in a variational sense and to directly apply it in the isogeometric framework. The elasticity problem in variational form, based on the principle of virtual work, reads

$$\int_\Omega \left(\mathbb{C}\nabla^S\mathbf{u}\right) : \nabla^S\mathbf{w}d\Omega = \int_\Omega \mathbf{f}\cdot\mathbf{w}d\Omega + \int_{\Gamma_p}\mathbf{p}\cdot\mathbf{w}d\Gamma \tag{21}$$

for every test function $\mathbf{w} \in \left[H^1(\Omega)\right]^{d_s}$ satisfying homogeneous Dirichlet boundary conditions, i.e.,

$$\mathbf{w} = \mathbf{0} \qquad \text{on}\,\Gamma_g \tag{22}$$

Integrating eq. (21) by parts and rearranging terms leads to

$$\int_\Omega \left[\nabla\cdot\left(\mathbb{C}\nabla^S\mathbf{u}\right) + \mathbf{f}\right]\cdot\mathbf{w}d\Omega - \int_{\Gamma_p}\left[\left(\mathbb{C}\nabla^S\mathbf{u}\right)\cdot\mathbf{n} - \mathbf{p}\right]\cdot\mathbf{w}d\Gamma = 0 \tag{23}$$

Note that, should the test function *not* satisfy eq. (22), we could release the Dirichlet boundary conditions on the solution $\mathbf{u}$ and introduce a Lagrange multiplier $\boldsymbol{\lambda} \in \left[H^{-1/2}(\Gamma_g)\right]^{d_s}$ to enforce the boundary conditions weakly. The variational form of the elasticity problem would then read: Find $(\mathbf{u},\boldsymbol{\lambda}) \in \left[H^1(\Omega)\right]^{d_s} \times \left[H^{-1/2}(\Gamma_g)\right]^{d_s}$ such that

$$\begin{aligned}\int_\Omega \left(\mathbb{C}\nabla^S\mathbf{u}\right) : \nabla^S\mathbf{w}d\Omega = \int_\Omega \mathbf{f}\cdot\mathbf{w}d\Omega + \int_{\Gamma_p}\mathbf{p}\cdot\mathbf{w}d\Gamma + \int_{\Gamma_g}\boldsymbol{\lambda}\cdot\mathbf{w}d\Gamma \\ + \int_{\Gamma_g}\boldsymbol{\mu}\cdot(\mathbf{u}-\mathbf{g})\,d\Gamma\end{aligned} \tag{24}$$

for every test function pair $(\mathbf{w},\boldsymbol{\mu}) \in \left[H^1(\Omega)\right]^{d_s} \times \left[H^{-1/2}(\Gamma_g)\right]^{d_s}$. Note that additional terms pertaining to the Dirichlet boundary have appeared above. If the solution $\mathbf{u}$ is sufficiently smooth (e.g., if $\mathbf{u} \in [H^2(\Omega)]^{d_s}$), then by elliptic regularity, it holds that $\left(\mathbb{C}\nabla^{\mathrm{S}}\mathbf{u}\right)\cdot\mathbf{n} \in \left[L^2(\Gamma_g)\right]^{d_s}$ and $\boldsymbol{\lambda} \equiv$

$(\mathbb{C}\nabla^S\mathbf{u}) \cdot \mathbf{n}$. Consequently, if $\mathbf{u}$ is sufficiently smooth, we have

$$\int_\Omega (\mathbb{C}\nabla^S\mathbf{u}) : \nabla^S\mathbf{w}d\Omega = \int_\Omega \mathbf{f}\cdot\mathbf{w}d\Omega + \int_{\Gamma_p} \mathbf{p}\cdot\mathbf{w}d\Gamma + \int_{\Gamma_g} \left[(\mathbb{C}\nabla^S\mathbf{u})\cdot\mathbf{n}\right]\cdot\mathbf{w}d\Gamma \tag{25}$$

for every weighting function $\mathbf{w} \in [H^1(\Omega)]^{d_s}$, and integration by parts leads to eq. (23). Therefore, if the solution is sufficiently smooth, the test function $\mathbf{w}$ does *not* need to satisfy homogeneous Dirichlet boundary conditions in order for eq. (23) to be applicable.

Using the isoparametric approach, we seek an approximation $\mathbf{u}^h$ to the unknown exact solution field $\mathbf{u}$ of the elastostatic problem in the form

$$\mathbf{u}^h = \sum_{a=1}^{n} R_a(\mathbf{x})\,\hat{\mathbf{u}}_a \tag{26}$$

where R_a are the NURBS basis functions described in Section 2.1 and representing the geometry of the problem, while $\hat{\mathbf{u}}_a$ are the unknown displacement control variables. Substitution into eq. (23) yields

$$\int_\Omega \left[\nabla\cdot(\mathbb{C}\nabla^S\mathbf{u}^h) + \mathbf{f}\right]\cdot\mathbf{w}d\Omega - \int_{\Gamma_p}\left[(\mathbb{C}\nabla^S\mathbf{u}^h)\cdot\mathbf{n} - \mathbf{p}\right]\cdot\mathbf{w}d\Gamma = 0 \tag{27}$$

We now need a suitable choice for the test function $\mathbf{w}$, which, in a collocation method, is selected as the Dirac delta, that can be formally constructed as the limit of a sequence of smooth functions with compact support converging to a distribution (Auricchio et al., 2010a, 2012a) and satisfying the so-called sifting property, i.e.,

$$\int_\Omega f_\Omega(\mathbf{x})\,\delta(\mathbf{x}-\mathbf{x}_i)\,d\Omega = f_\Omega(\mathbf{x}_i) \tag{28}$$

$$\int_\Gamma f_\Gamma(\mathbf{x})\,\delta(\mathbf{x}-\mathbf{x}_i)\,d\Gamma = f_\Gamma(\mathbf{x}_i) \tag{29}$$

for every function f_Ω continuous about the point $\mathbf{x}_i \in \Omega$ and for every function f_Γ continuous about the point $\mathbf{x}_i \in \Gamma$. In the following, the Dirac delta will be indicated as a Dirac delta "function" following conventional terminology.

Let us assume that $d_s = 2$, m_1 and m_2 are the numbers of control points in the two parametric directions, and $n = m_1 m_2$ is the total number of control points. Thus $2n$ scalar equations are needed to determine the unknown displacement control variables. Analogously to the previous section, we choose n collocation points τ_{kl}, $k = \{1, ..., m_1\}$, $l = \{1, ..., m_2\}$ located at the images of the (tensor product) Greville abscissae of the knot vectors. The collocation points for $k = 1, m_1$ and $l = 1, m_2$ are located at the boundary Γ. Separate sets of equations are needed for the patch interior and for the boundaries.

In the patch interior Ω, we write $2(m_1 - 2)(m_2 - 2)$ scalar equations by choosing as test functions the Dirac delta functions centered at the interior collocation points τ_{kl}, $k = \{2, ..., m_1 - 1\}$, $l = \{2, ..., m_2 - 1\}$. The resulting equations read

$$\left[\nabla \cdot \left(\mathbb{C}\nabla^S \mathbf{u}^h\right) + \mathbf{f}\right](\tau_{kl}) = \mathbf{0} \qquad \tau_{kl} \subset \Omega \tag{30}$$

i.e., they are the collocated strong form of the equations at τ_{kl}.

No equations are needed at the Dirichlet boundary, as we impose *a priori* that $u_i^h(\tau_{kl}) = g_i(\tau_{kl})$ on Γ_g.

To enforce Neumann boundary conditions, each $\tau_{kl} \subset \Gamma_p$ is associated with a collocation equation that sets the value of the boundary traction. This corresponds to choosing as test functions the Dirac delta functions centered at the collocation points located at the Neumann boundary. Here a distinction is needed between the collocation points located at the edges ($k = 1, m_1$ and $l = 2, ..., m_2 - 1$, or $l = 1, m_2$ and $k = 2, ..., m_1 - 1$), and those located at the corners of the domain ($k = 1, m_1$ and $l = 1, m_2$). For collocation points located on edges within the Neumann boundary, the equations are

$$\left[\left(\mathbb{C}\nabla^S \mathbf{u}^h\right) \cdot \mathbf{n} - \mathbf{p}\right](\tau_{kl}) = \mathbf{0} \qquad \tau_{kl} \subset \text{edge} \subset \Gamma_p \tag{31}$$

i.e., they are the collocated strong form of the Neumann boundary conditions at τ_{kl}. For collocation points located at corners where two Neumann boundaries meet, it has been shown in Auricchio et al. (2012a)[3] that the appropriate equations are

$$\left[\left(\mathbb{C}\nabla^S \mathbf{u}^h\right) \cdot \mathbf{n}' - \mathbf{p}'\right](\tau_{kl}) + \left[\left(\mathbb{C}\nabla^S \mathbf{u}^h\right) \cdot \mathbf{n}'' - \mathbf{p}''\right](\tau_{kl}) = \mathbf{0} \quad \tau_{kl} \equiv \text{corner} \subset \Gamma_p \tag{32}$$

where $\mathbf{n}'$ and $\mathbf{n}''$ are the unit outward normals of the edges meeting at the corner, and $\mathbf{p}'$ and $\mathbf{p}''$ are the respective imposed tractions.

It is important to remark that, in a recent paper by De Lorenzis et al. (2014b), it has been shown that the above approach to impose Neumann boundary conditions may lead to spurious oscillations in cases implying solutions of reduced regularity, when non-uniform meshes are adopted. There-

[3] In addition, interested readers may find in Auricchio et al. (2012a) a detailed discussion on the conditions to be imposed in more complicate situations like at the interfaces of multi-patch geometries.

fore alternative methods are needed, and in the same paper two possible (simple) alternative strategies to cure this issue are presented.

3.3 Numerical results.

We now present some numerical experiments showing the behavior of the proposed collocation formulation. In particular, we propose: i) a plane strain test, to show the fact that the convergence rates observed in 1D are attained also on mapped geometries in higher dimensions; ii) a two-material two-patch plane strain traction test, to show the fact that constant strain states are represented exactly (also in the case of multi-patch analysis); and iii) a 3D test on an elastic block, on which a performance comparison among IGA collocation and Galerkin IGA and FEA is carried out.

Plane strain clamped quarter of an annulus. As a first test, we consider a plane strain quarter of an annulus, as sketched in Figure 7, with internal and external radii equal to $R_1 = 1$ and $R_2 = 4$, respectively. The domain can be exactly represented by a single quadratic NURBS patch.

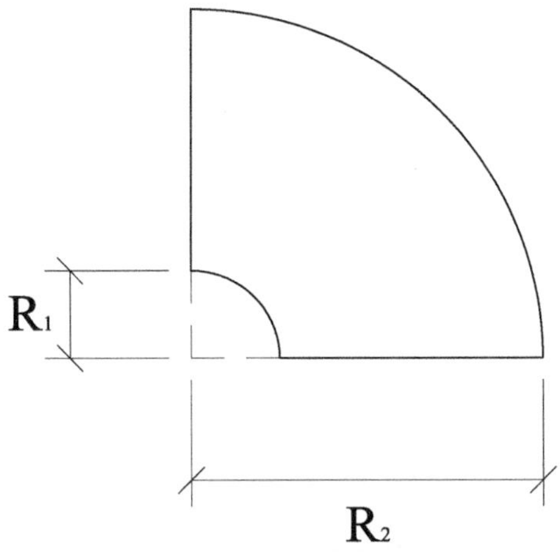

Figure 7. Plane strain clamped quarter of an annulus. Problem geometry.

The whole domain boundary is assumed to be clamped and we assign a manufactured solution in terms of displacement components, reading:

$$\begin{cases} u = (x^2 + y^2 - 1)(x^2 + y^2 - 16)\sin(x)\sin(y), \\ v = (x^2 + y^2 - 1)(x^2 + y^2 - 16)\sin(x)\sin(y). \end{cases} \tag{33}$$

The manufactured solution satisfies the prescribed boundary conditions and the load is computed from it by imposing equilibrium.

The problem is solved by IGA collocation using Greville abscissae and in Figure 8 we present the results in terms of relative solution error in the L^2-norm versus the square root of the total number of control points. It is possible to observe that the convergence rates observed in the 1D case (i.e., p and $p-1$ for even and odd degree p, respectively) are attained.

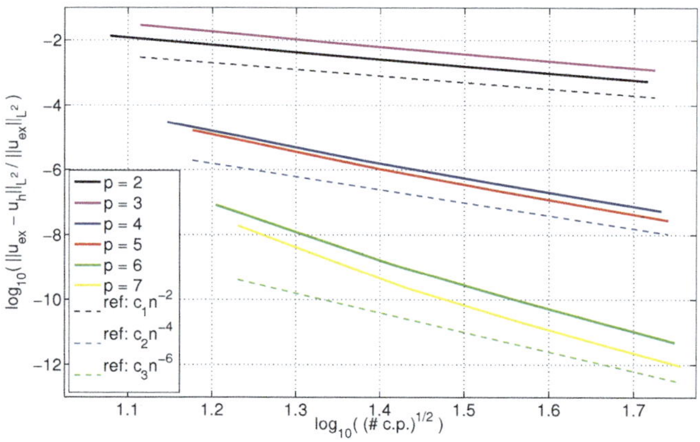

Figure 8. Plane strain clamped quarter of an annulus. Error plot versus the square root of number of control points for different degree NURBS.

Two-material two-patch plane strain traction test. We now consider a rectangular domain Ω, as sketched in Figure 9, subjected to a uniform traction. The domain is assumed to consist of two material subdomains.

The idea is to reproduce a solution homogeneous in the y direction (and piece-wise homogeneous in the x direction), such that the numerical results should be able to exactly reproduce the analytical solution. To obtain such a solution, it is necessary to properly calibrate the elastic constants. Accordingly, considering the standard plane strain equations for each material,

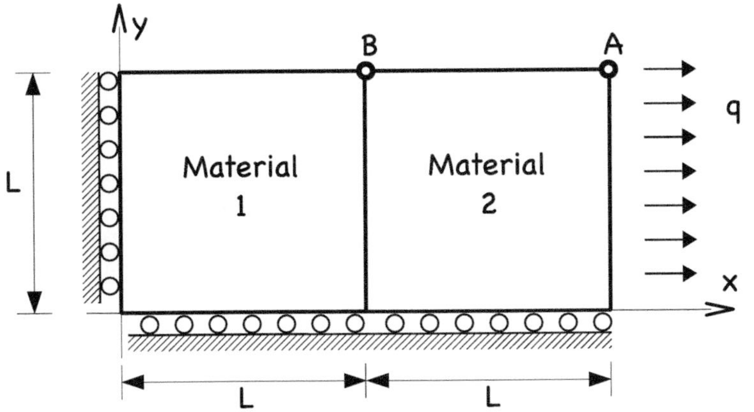

Figure 9. Two-material two-patch plane strain traction test. Problem geometry and boundary conditions.

we require the transverse strain (i.e., the strain in the transverse direction with respect to the traction direction) to be the same in both materials, obtaining the following relation

$$\frac{E_1}{E_2} = \frac{\nu_1 (1 + \nu_1)}{\nu_2 (1 + \nu_2)}$$

where the subscripts indicate the material numbers.

The problem under investigation is of particular interest also because it presents several noteworthy boundary conditions situations (e.g., it introduces a boundary point – point B in Figure 9 – with a combination of a traction-free boundary condition and an interface between different materials).

As material properties we assume $\nu_1 = 0.2$, $\nu_2 = 0.25$, and $E_2 = 1000$, resulting in $E_1 = 768$. With these choices, the exact displacement of point A (indicated in Figure 9) can be easily computed to be

$$u_A = 2.1875 \cdot 10^{-3} \quad v_A = 3.125 \cdot 10^{-4}$$

We numerically solve the problem by IGA collocation using two conforming NURBS patches (i.e., a patch for each material). The analytical solution is matched up to machine precision by the numerical one computed using a single element per patch, illustrating the good behavior of the proposed numerical scheme for the case under investigation.

Figure 10 shows the horizontal and vertical displacement fields (obtained using $p = q = 2$ and 3×3 control points per patch, i.e., one element per patch), which are linear in the two coordinate variables within each material, as expected. We also highlight that, as desired, a perfectly homogeneous solution is obtained in the y direction.

We may also notice that the management of a conforming multi-patch situation is very simple in the proposed collocation method, since it is based on constructing the discrete equilibrium relation for each patch and, then, summing the equations associated to collocation points shared by multiple patches (see Auricchio et al., 2012a, for more details).

3D elasticity test and comparison with Galerkin. We finally present the results of a 3D test on an elastic block, focusing in particular our attention on performance comparison among IGA collocation and Galerkin IGA and FEA. These results have been originally proposed by Schillinger et al. (2013), where a comprehensive study on the computational costs of IGA collocation as well as of Galerkin IGA and FEA is reported, including detailed operation counts showing the potential advantages of IGA collocation. Readers are therefore referred to such a paper for more details.

The considered problem consists of an elastic cube defined over the domain $[0,1]^3$, fully clamped on its boundary and loaded by a body force giving rise to a manufactured solution $u_1 = u_2 = u_3 = \sin(2\pi x_1)\sin(2\pi x_2)\sin(2\pi x_3)$. Material parameters are selected to be $E = 1$ and $\nu = 0.3$.

Numerical results in terms of L^2- and H^1-norm relative errors versus number of degrees of freedom per parametric direction are reported in Figures 11 and 12, respectively. It is possible to see that, for IGA collocation, the same convergence rates observed in 1D are attained. For Galerkin IGA and FEA, instead, optimal convergence rates are attained, as expected from the theory. In terms of convergence constants, the best results are always guaranteed by Galerkin IGA, whereas IGA collocation presents slightly worse or similar results with respect to Galerkin FEA in the L^2-norm, and better results in the H^1-seminorm (except for the cubic case that, interestingly enough, in collocation is always underachieving, in particular with respect to the quadratic case).

Having in mind accuracy-to-computational-cost ratio as a measure of the efficiency of a numerical method, we also report results in terms of L^2- and H^1-norm relative errors versus computational times. The obtained results, reported in Figures 13 and 14, show that for degrees $p > 3$ IGA collocation guarantees the best overall performance, heavily outperforming Galerkin approaches in the H^1-seminorm. Such results are also supported by the operation counts provided in Schillinger et al. (2013). In particular,

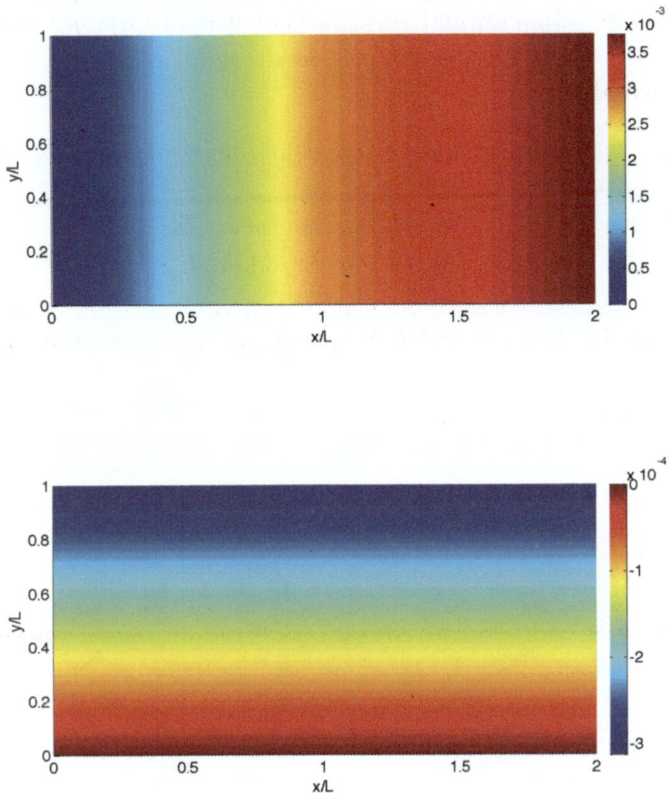

Figure 10. Two-material two-patch plane strain traction test. Horizontal (top) and vertical (bottom) displacement fields.

it is worth to remark that, as highlighted in the above-mentioned paper, the superiority of IGA collocation is best illustrated in Figure 14, where with $p = 4$ IGA collocation achieves an error level of 10^{-5} in the H^1-seminorm in less than 20 seconds, whereas both Galerkin approaches require more than 500 seconds to reach that accuracy.

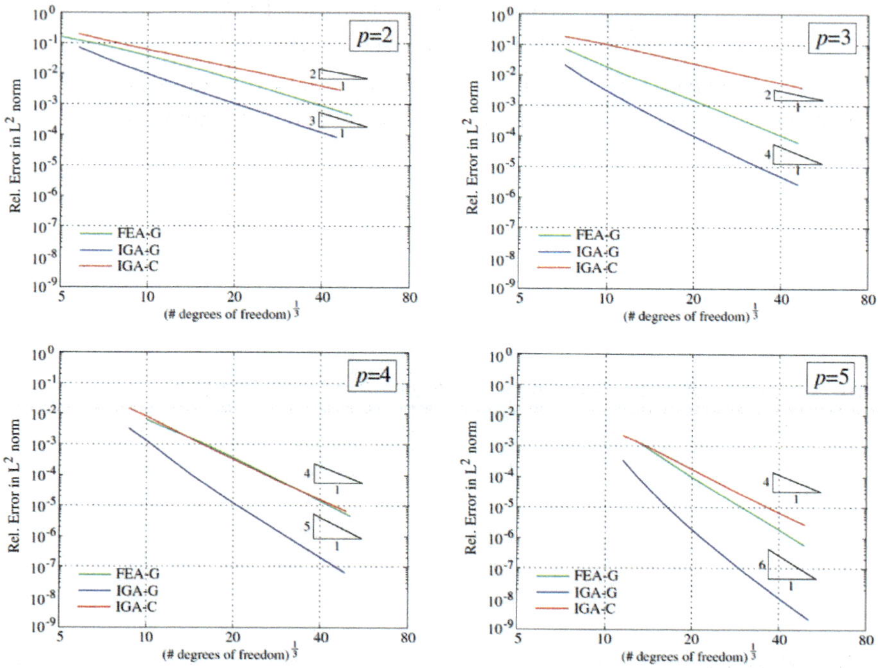

Figure 11. 3D elasticity test. Relative error in L^2-norm versus number of degrees of freedom per parametric direction for IGA collocation (IGA-C) as well as Galerkin IGA (IGA-G) and FEA (FEA-G).

4 A Brief Overview on Other Results and Applications

In this section, we give a brief overview on further results and interesting applications in the field of IGA collocation. Since the aim of the present chapter is just to give a basic and concise introduction to IGA collocation,

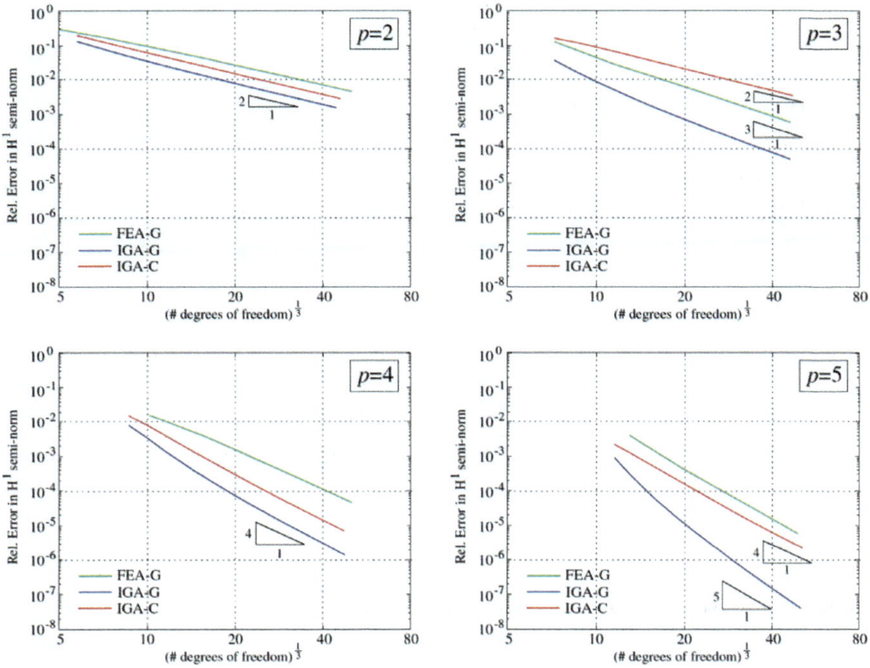

Figure 12. 3D elasticity test. Relative error in H^2-seminorm versus number of degrees of freedom per parametric direction for IGA collocation (IGA-C) as well as Galerkin IGA (IGA-G) and FEA (FEA-G).

this section is not meant as a complete and self-consistent review, and readers are referred to the cited individual papers for more information and details.

Explicit dynamics. Probably, the most promising application of IGA collocation is explicit dynamics. In fact, in explicit dynamics analyses, the speed is almost entirely dependent on the cost of quadrature, because the computational cost is dominated by stress divergence evaluations at quadrature points for the calculation of the residual force vector. Accordingly, explicit codes used extensively for crash dynamics and metal forming, such as LS-DYNA, rely almost exclusively on fast low-order quadrilateral and hexahedral elements with one-point quadrature. Unfortunately, those elements

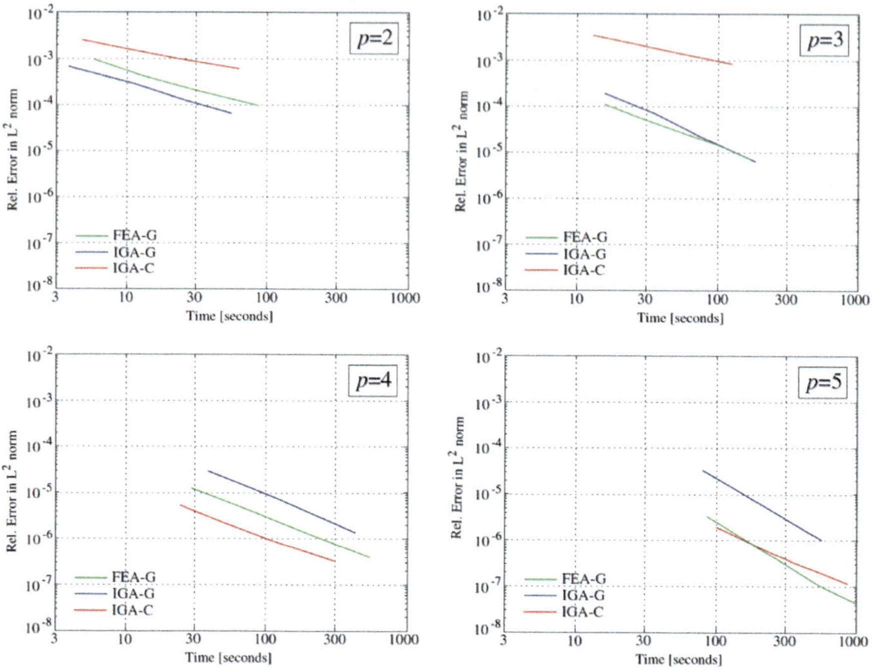

Figure 13. 3D elasticity test. Relative error in L^2-norm versus computational time for IGA collocation (IGA-C) as well as Galerkin IGA (IGA-G) and FEA (FEA-G).

typically require stabilization (whose parameters usually need fine-tuning by computationally expensive and time-consuming sensitivity studies) against mesh instabilities such as hourglass modes.

In this context, isogeometric collocation can be viewed as a one-point quadrature scheme that is rank sufficient and is therefore free of mesh instabilities. Hence, IGA collocation methods eliminate the need for ad hoc hourglass stabilization techniques and the relative parameter tunings, and can be seen as a very promising fast and accurate higher-order alternative (further considerations on the advantages of IGA collocation in explicit dynamics can be found in Auricchio et al., 2012a, and Schillinger et al., 2013). In addition, IGA collocation methods show great promise for the development of higher-order accurate time integration schemes due to the

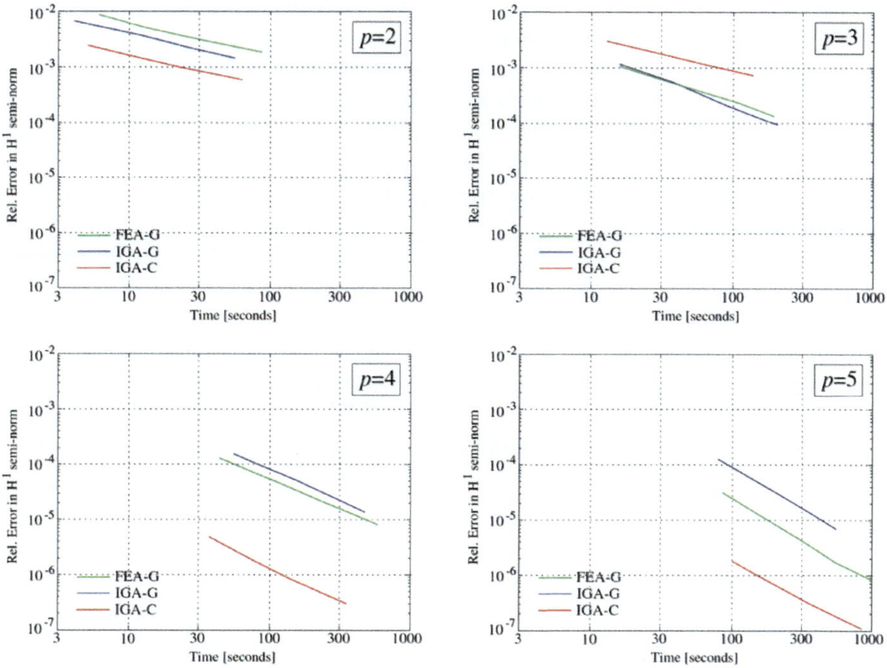

Figure 14. 3D elasticity test. Relative error in H^2-seminorm versus computational time for IGA collocation (IGA-C) as well as Galerkin IGA (IGA-G) and FEA (FEA-G).

convergence of the high modes in the eigenspectrum (cf. the spectral results in Section 2.2).

However, as in IGA Galerkin, classical row-sum mass lumping techniques limit accuracy of standard explicit schemes to second order only, independently of the approximation degree (Cottrell et al., 2006). In Auricchio et al. (2012a), a predictor-multicorrector scheme has been proposed to conveniently recover higher-order convergence rates in space within an explicit framework. Such a method is however only second-order in time. Moving from four-stage Runge-Kutta integration methods, similar results have been obtained also within a fourth order scheme in time (and preliminary results have been recently presented in Reali et al., 2014).

Local refinement. As we have seen, standard NURBS-based IGA collocation is based on tensor-product NURBS, which are used within the isoparametric paradigm for the representation of both geometry and field variables. Once the (tensor-product) geometry representation is fixed, also the distribution of the collocation points (e.g., Greville abscissae) is fixed and affected by the tensor-product structure. Therefore, when local refinement is needed, it is necessary to resort to alternative locally-refinable shape functions.

In Schillinger et al. (2013), this has been achieved resorting to hierarchical B-Splines and NURBS. However, in that paper, it has been shown that the plain use of those shape functions is not enough to get a working locally-refinable method, because linear independence problems may arise in transition regions between different levels. The proposed solution is to use hierarchical NURBS spaces in combination with the concept of weighted IGA collocation (practically applying the weighting concept only in transition regions). The resulting method is simple and efficient, and an accurate description can be found in Schillinger et al. (2013) along with several numerical tests proving its effectiveness. Among them, one of the most significant ones is a 3D advection-diffusion benchmark consisting of a cylinder that rotates around its axis with a tangential velocity $a_\theta = \omega r$ and a radial velocity $a_r = 0$, with a flow of constant axial velocity a_z, which results in a helical plume of the concentration that emerges from the fixed local inflow boundary condition $u = 1$. In Figure 15, we report the problem definition, the adopted hierarchical mesh (defined by an automatic adaptive refinement procedure based on a gradient-based error indicator), and the corresponding solution. A uniform discretization that yields a plume resolution with the same small element size as in the adaptive mesh requires a globally refined mesh with 1,095,200 degrees of freedom, whereas the presented adaptive mesh requires only 104,017 degrees of freedom.

Contact. When dealing with the simulation of practical structural mechanics problems involving interactions between multiple patches with nonconforming discretizations, the enforcement of contact constraints is a fundamental issue to be properly addressed. IGA contact is now an established research field in the Galerkin context, where it has been shown that the higher continuity typical of IGA shape functions can have beneficial effects (see, e.g., the recent review by De Lorenzis et al., 2014c, and references therein). More recently, the problem of frictionless contact has been successfully tackled also in the framework of IGA collocation by De Lorenzis et al. (2014b), who have proposed a two-half-pass algorithm. In that paper, several demanding examples have been considered in order to prove the

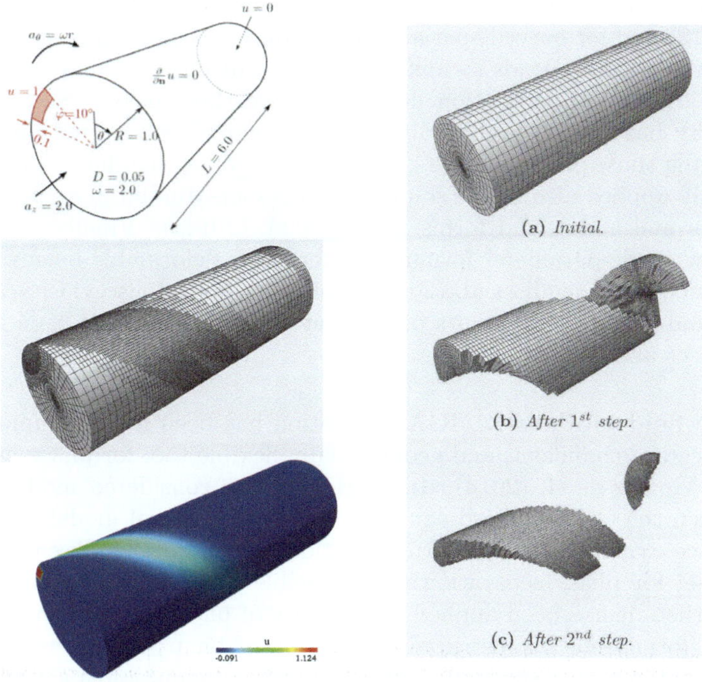

(a) *Initial.*

(b) *After* 1st *step.*

(c) *After* 2nd *step.*

Figure 15. Advection-diffusion in a rotating cylinder with hierarchical weighted collocation. Problem definition (top left); adopted hierarchical mesh (middle left); solution (bottom left); and finest elements at the different steps of the adaptive procedure (right).

good behavior of IGA collocation in contact problems. In particular, we mention here the remarkable fact that the contact patch test is passed up to machine precision with IGA collocation, something that does not happen for standard node-to-surface algorithms with Galerkin formulations.

Structural elements. The peculiar features of IGA collocation has been also widely applied for the development of structural thin elements like beams and plates. In particular, Bernoulli-Euler beam and Kirchhoff plate elements have been proposed by Reali and Gomez (2014). Also shear-deformable structural elements have been considered and, in the case of three-field (i.e., displacements, rotations, shear stresses) mixed formulations for Timoshenko initially-straight planar beams (Beirão da Veiga et al., 2012) and for curved spatial rods (Auricchio et al., 2013), the structure of IGA collocation leads to mixed methods which are locking-free independently of the approximation degrees for the three fields. Such a unique property has been proven analytically and extensively tested numerically. Following those positive results, isogeometric collocation has been then successfully applied also to the solution of Reissner-Mindlin plate problems, in both primal and mixed forms Kiendl et al. (2014a). Finally, an interesting new single-parameter formulation for shear-deformable beams, recently introduced by Kiendl et al. (2014b), has been solved also via IGA collocation, and preliminary results on its plate counterpart have been shown in Kiendl et al. (2014c).

Phase-field modeling. IGA collocation has been finally employed as a fast, accurate, efficient, and geometrically flexible tool for phase-field models by Gomez et al. (2014). In particular, the considered model problem has been the Cahn-Hilliard equation, which is a central model in nonlinear interface dynamics and pattern formation, derived about fifty years ago as a model for phase separation of immiscible fluids by Cahn and Hilliard. Since then, it has been applied to a variety of physical problems, including planet formation, microstructure evolution of binary mixtures, and phase separation of polymer blend. The Cahn-Hilliard equation is also one of the simplest equations that can model stable co-existence of two phases and, as such, is the basis for various multiphase flow theories (interested readers are referred to the related references cited in Gomez et al., 2008, 2014). In the context of the numerical solution of the Cahn-Hiliard equation, IGA collocation proposes itself as a successful combination of the geometrical flexibility of Galerkin FEA and IGA approaches and the accuracy, efficiency, and simplicity of pseudo-spectral collocation methods. Several significant numerical examples are reported in Gomez et al. (2014), and two of them are shown in Figure 16. Given these positive results and the proven potential of phase-field modeling within the IGA framework (see, e.g., Gomez et al., 2008; Borden et al., 2012), this appears to be one of the most promising fields for the implementation of IGA collocation, in particular for what concerns the context of fracture mechanics (cf. De Lorenzis et al., 2014a, and Schillinger et al., 2014b).

5 Conclusions and further current developments.

The aim of this chapter was to give a concise introduction to the recently introduced and more than promising family of isogeometric methods based

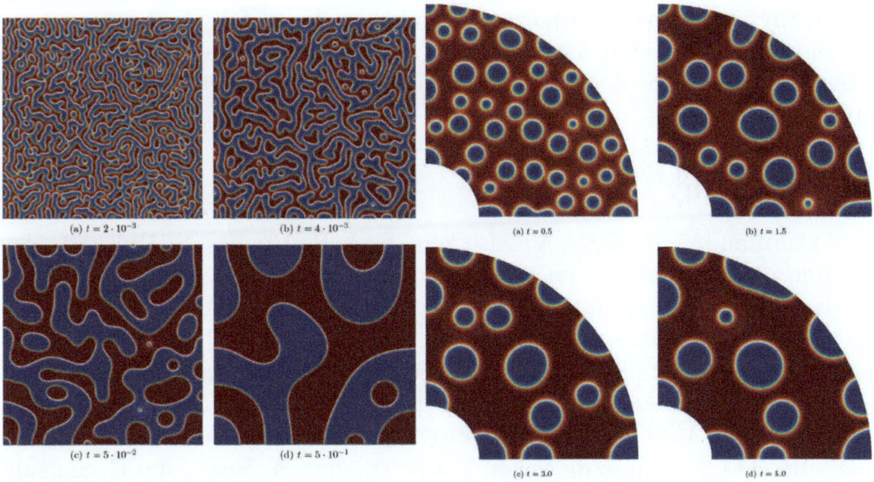

Figure 16. Phase-field modeling via IGA collocation. Spinodal decomposition on a square domain (left) and nucleation on a mapped domain (right). See Gomez et al. (2014) for details and parameter values.

on the collocation concept. Accordingly, the basic ideas of the approach have been first introduced in 1D along with some theoretical results, and then extended to higher dimensions in the framework of linear elastostatics problems. Moreover, several significant numerical tests have been shown, in order to confirm the good behavior and potential of the method, also on the basis of the accuracy-to-computational-cost ratio and in comparison with Galerkin approaches.

We have finally given a brief overview of further results and interesting applications of IGA collocation, including local refinement, contact, structural elements, and phase-field modeling. In addition to the above-mentioned applications, some promising preliminary results have been also recently obtained in dealing with the incompressibility constraint via mixed methods (Morganti et al., 2014), in the nonlinear elastic and inelastic regimes (De Lorenzis et al., 2014a), as well as in the field of Computational Fluid Dynamics (Evans et al., 2013).

As it can be seen, due to its special features, IGA collocation has shown so far a lot of potential in many fields of Computational Mechanics and is attracting more and more attention among researchers. However, there are many issues that still have to be addressed and fully understood (e.g., a complete and sound mathematical analysis of the method is yet to be developed), and the door is definitely open for many further improvements and applications.

Bibliography

I. Akkerman, Y. Bazilevs, V. M. Calo, T. J. R. Hughes, S. Hulshoff. The role of continuity in residual-based variational multiscale modeling of turbulence. *Computational Mechanics*, 41:371–378, 2007.

F. Auricchio, L. Beirão da Veiga, A. Buffa, C. Lovadina, A. Reali, and G. Sangalli. A fully "locking-free" isogeometric approach for plane linear elasticity problems: A stream function formulation. *Computer Methods in Applied Mechanics and Engineering*, 197:160–172, 2007.

F. Auricchio, L. Beirão da Veiga, T.J.R. Hughes, A. Reali, and G. Sangalli. Isogeometric collocation methods. *Mathematical Models and Methods in Applied Sciences*, 20(11):2075–2107, 2010.

F. Auricchio, L. Beirão da Veiga, T.J.R. Hughes, A. Reali, and G. Sangalli. Isogeometric collocation for elastostatics and explicit dynamics. *Computer Methods in Applied Mechanics and Engineering*, 249-252:2–14, 2012.

F. Auricchio, L. Beirão da Veiga, J. Kiendl, C. Lovadina, and A. Reali. Locking-free isogeometric collocation methods for spatial Timoshenko rods. *Computer Methods in Applied Mechanics and Engineering*, 263:113–126, 2013.

F. Auricchio, L. Beirão da Veiga, C. Lovadina, A. Reali. The importance of the exact satisfaction of the incompressibility constraint in nonlinear elasticity: mixed FEMs versus NURBS-based approximations. *Computer Methods in Applied Mechanics and Engineering*,199:314–323, 2010.

F. Auricchio, F. Calabrò, T.J.R. Hughes, A. Reali, and G. Sangalli. A simple algorithm for obtaining nearly optimal quadrature rules for NURBS-based isogeometric analysis. *Computer Methods in Applied Mechanics and Engineering*, 249–252:15–27, 2012.

Y. Bazilevs, V.M. Calo, J.A. Cottrell, T.J.R. Hughes, A. Reali, G. Scovazzi. Variational multiscale residual-based turbulence modeling for large eddy simulation of incompressible flows. *Computer Methods in Applied Mechanics and Engineering*, 197:173–201, 2007.

Y. Bazilevs, T.J.R. Hughes. NURBS-based isogeometric analysis for the computation of flows about rotating components. *Computational Mechanics*, 43:143–150, 2008.

L. Beirão da Veiga, T.J.R. Hughes, J. Kiendl, C. Lovadina, J. Niiranen, A. Reali, and H. Speelers. A locking-free model for Reissner-Mindlin plates: Analysis and isogeometric implementation via NURBS and triangular NURPS. *In preparation*, 2014.

L. Beirão da Veiga, C. Lovadina, and A. Reali. Avoiding shear locking for the Timoshenko beam problem via isogeometric collocation methods. *Computer Methods in Applied Mechanics and Engineering*, 241-244:38–51, 2012.

M.J. Borden, C.V. Verhoosel, M.A. Scott, T.J.R. Hughes, C.M. Landis. A phase-field description of dynamic brittle fracture. *Computer Methods in Applied Mechanics and Engineering*, 217:77–95, 2012.

J.F. Caseiro, R.A.F. Valente, A. Reali, J. Kiendl, F. Auricchio, R.J. Alves de Sousa. On the Assumed Natural Strain method to alleviate locking in solid-shell NURBS-based finite elements. *Computational Mechanics*, 53:1341–1353, 2014.

J.A. Cottrell, T.J.R. Hughes, Y. Bazilevs. *Isogeometric Analysis: Toward Integration of CAD and FEA*. Wiley, 2009.

J.A. Cottrell, T.J.R. Hughes, A. Reali. Studies of refinement and continuity in isogeometric structural analysis. *Computer Methods in Applied Mechanics and Engineering*,196:4160–4183, 2007.

J.A. Cottrell, A. Reali, Y. Bazilevs, and T.J.R. Hughes. Isogeometric analysis of structural vibrations. *Computer Methods in Applied Mechanics and Engineering*, 195:5257–5296, 2006.

A. Buffa, C. de Falco, G. Sangalli. IsoGeometric Analysis: Stable elements for the 2D Stokes equation. *International Journal for Numerical Methods in Fluids*, 65:1407–1422, 2011.

C. de Boor. *A practical guide to Splines (revised edition)*. Springer, 2001.

C. de Falco, A. Reali, R. Vázquez. GeoPDEs: a research tool for IsoGeometric Analysis of PDEs. *Advances in Engineering Software*, 42:1020–1034, 2011.

L. De Lorenzis, M. Ambati, J.A. Evans, T.J.R. Hughes, R. Kruse, N. Nguyen-Thanh, A. Reali. *Recent results on isogeometric collocation for the solution of non-linear problems*. Invited lecture at HOFEIM 2014 – Workshop on Higher Order Finite Element and Isogeometric Methods, Munich, July 15-18, 2014.

L. De Lorenzis, J.A. Evans, T.J.R. Hughes, A. Reali. Isogeometric collocation: Neumann boundary conditions and contact. *Computer Methods in Applied Mechanics and Engineering*, doi:10.1016/j.cma.2014.06.037, 2014.

L. De Lorenzis, T.J.R. Hughes, P. Wriggers. Isogeometric contact: a review. *GAMM MItteilungen*, 37:85–123, 2014.

S. Demko. On the existence of interpolation projectors onto spline spaces, *Journal of Approximation Theory*, 43:151–156, 1985.

R.P. Dhote, H. Gomez, R N.V. Melnik, J. Zu. Isogeometric analysis of a dynamic thermo-mechanical phase-field model applied to shape memory alloys. *Computational Mechanics*, 53:1235–1250, 2014.

T. Elguedj, Y. Bazilevs, V.M. Calo, T.J.R. Hughes. $\bar{B}$ and $\bar{F}$ projection methods for nearly incompressible linear and non-linear elasticity and plasticity using higher-order NURBS elements. *Computer Methods in Applied Mechanics and Engineering*, 197:2732–2762, 2008.

J.A. Evans, R. Hiemstra, T.J.R. Hughes, D. Schillinger. *Isogeometric Divergence-Conforming Collocation Methods for Incompressible Fluid Flow.* Invited lecture at the 12th U.S. National Congress on Computational Mechanics, Raleigh, July 22-25, 2013.

H. Gomez, V. M. Calo, Y. Bazilevs, and T. J. R. Hughes. Isogeometric analysis of the Cahn–Hilliard phase-field model. *Computer Methods in Applied Mechanics and Engineering*, 197:4333–4352, 2008.

H. Gomez, T.J.R. Hughes, X. Nogueira, V.M. Calo. Isogeometric analysis of the isothermal Navier-Stokes-Korteweg equations. *Computer Methods in Applied Mechanics and Engineering*, 199:1828–1840, 2010.

H. Gomez, A. Reali, and G. Sangalli. Accurate, efficient, and (iso)geometrically flexible collocation methods for phase-field models. *Journal for Computational Physics*, 262:153–171, 2014.

T.J.R. Hughes, J.A. Cottrell, and Y. Bazilevs. Isogeometric analysis: CAD, finite elements, NURBS, exact geometry, and mesh refinement. *Computer Methods in Applied Mechanics and Engineering*, 194:4135–4195, 2005.

T.J.R. Hughes, J.A. Evans, A. Reali. Finite Element and NURBS Approximations of Eigen- value, Boundary-value, and Initial-value Problems. *Computer Methods in Applied Mechanics and Engineering*, 272:290–320, 2014.

T.J.R. Hughes, A. Reali, and G. Sangalli. Duality and unified analysis of discrete approximations in structural dynamics and wave propagation: Comparison of p-method finite elements with k-method NURBS. *Computer Methods in Applied Mechanics and Engineering*, 197:4104–4124, 2008.

T.J.R. Hughes, A. Reali, and G. Sangalli. Efficient quadrature for NURBS-based isogeometric analysis. *Computer Methods in Applied Mechanics and Engineering*, 199:301–313, 2010.

J. Kiendl, F. Auricchio, L. Beirao da Veiga, C. Lovadina, A. Reali. Isogeometric collocation methods for the Reissner-Mindlin plate problem. Submitted, 2014.

J. Kiendl, F. Auricchio, T.J.R. Hughes, A. Reali. Single-variable formulations and isogeometric discretizations for shear deformable beams. ICES Report 14-26, 2014.

J. Kiendl, F. Auricchio, T.J.R. Hughes, A. Reali. *Isogeometric one-parameter formulations for shear deformable structures.* Keynote lecture at WCCM-ECCM-ECFD2014 – 11th World Congress on Computational Mechanics, 5th European Conference on Computational Methods, and 6th European Conference on Computational Fluid Dynamics, Barcelona, July 20-25, 2014.

J. Kiendl, K.-U. Bletzinger, J. Linhard, and R. Wüchner. Isogeometric shell analysis with Kirchhoff-Love elements. *Computer Methods in Applied Mechanics and Engineering*, 198:3902–3914, 2009.

S. Lipton, J.A. Evans, Y. Bazilevs, T. Elguedj, T.J.R. Hughes. Robustness of isogeometric structural discretizations under severe mesh distortion. *Computer Methods in Applied Mechanics and Engineering*, 199:357–373, 2010.

J. Liu, H. Gomez, J.A. Evans, T.J.R. Hughes, C.M. Landis. Functional entropy variables: A new methodology for deriving thermodynamically consistent algorithms for complex fluids, with particular reference to the isothermal Navier-Stokes-Korteweg equations. new block *Journal of Computational Physics*, 248:47–86, 2014.

S. Morganti, F. Auricchio, D.J. Benson, F.I. Gambarin, S. Hartmann, T.J.R. Hughes, A. Reali. Patient-specific isogeometric structural analysis of aortic valve closure. *ICES Report*, 14-10, 2014.

S. Morganti, F. Auricchio, L. De Lorenzis, J.A. Evans, T.J.R. Hughes, A. Reali. *Isogeo- metric collocation: incompressible elasticity, locking and possible solutions.* Invited lecture at WCCM-ECCM-ECFD2014 – 11th World Congress on Computational Mechanics, 5th European Conference on Computational Methods, and 6th European Conference on Computational Fluid Dynamics, Barcelona, July 20-25, 2014.

L. Piegl, W. Tiller. The NURBS Book, 2nd Edition. Springer-Verlag, 1997.

A. Reali. An isogeometric analysis approach for the study of structural vibrations. *Computer Methods in Applied Mechanics and Engineering*, 1–30:15–27, 2006.

A. Reali, H. Gomez. An isogeometric collocation approach for Bernoulli-Euler beams and Kirchhoff plates. *Submitted*, 2014.

A. Reali, J.A. Evans, T.J.R. Hughes. *Isogeometric Analysis: Structural vibrations and dynamics.* Semi-plenary lecture at WCCM-ECCM-ECFD2014 – 11th World Congress on Computational Mechanics, 5th European Conference on Computational Methods, and 6th European Conference on Computational Fluid Dynamics, Barcelona, July 20-25, 2014.

D. Schillinger, J.A. Evans, A. Reali, M.A. Scott, T.J.R. Hughes. Isogeometric collocation: Cost comparison with Galerkin methods and extension to adaptive hierarchical NURBS discretizations. *Computer Methods in Applied Mechanics and Engineering*, 267:170–232, 2013.

D. Schillinger, S.J. Hossain, T.J.R. Hughes. Reduced Bézier element quadrature rules for quadratic and cubic splines in isogeometric analysis. *Computer Methods in Applied Mechanics and Engineering*, 277:1–45, 2014.

D. Schillinger, J.A. Evans, T.J.R. Hughes. *Cost of collocation, accuracy of Galerkin: On the potential of higher-order collocation-type methods in IGA and hp FEM.* Invited lecture at HOFEIM 2014 – Workshop on Higher Order Finite Element and Isogeometric Methods, Munich, July 15-18, 2014.

Isogeometric analysis based on T-splines

Derek C. Thomas[*] and Michael A. Scott [†]

[*] Department of Physics and Astronomy
Brigham Young University
[†] Department of Civil and Environmental Engineering
Brigham Young University
[‡] common affil

Abstract This chapter provides an introduction to the use of T-splines in isogeometric analysis. A simple definition of two-dimensional T-splines is given and Bézier extraction is introduced. The basic details for implementation of T-splines as finite element shape functions are given. Two examples of integrated analysis and design based on commercial tools are given to illustrate the utility of T-spline-based IGA in a design-through-analysis workflow.

1 Introduction

Isogeometric analysis (Hughes et al. (2005); Cottrell et al. (2009)) is a generalization of finite element analysis which improves the link between geometric design and analysis. The isogeometric paradigm is simple: the smooth spline basis used to define the geometry is used as the basis for analysis. As a result, exact geometry is introduced into the analysis. The smooth basis can be leveraged by the analysis (Evans et al. (2009); Hughes et al. (2014); Cottrell et al. (2007)) leading to innovative approaches to model design (Cohen et al. (2010); Wang et al. (2011); Liu et al. (2014)), analysis (Schillinger et al. (2012); Scott et al. (2013); Schmidt et al. (2012); Benson et al. (2010a)), optimization (Wall et al. (2008)), and adaptivity (Bazilevs et al. (2010); Dörfel et al. (2009); Scott et al. (2014); Evans et al. (2014)).

Many of the early isogeometric developments were restricted to NURBS but the use of T-splines as an isogeometric basis has gained widespread attention across a number of application areas Bazilevs et al. (2010); Scott et al. (2011, 2012); Verhoosel et al. (2011b,a); Borden et al. (2012); Benson et al. (2010a); Schillinger et al. (2012); Scott et al. (2013); Simpson et al. (2014); Dimitri et al. (2014); Hosseini et al. (2014); Bazilevs et al. (2012); Buffa et al. (2014); Ginnis et al. (2014). Particular focus has been placed

G. Beer, S. Bordas (Eds.), *Isogeometric Methods for Numerical Simulation*,
CISM International Centre for Mechanical Sciences
DOI 10.1007/ 978-3-7091-1843-6_5 © CISM Udine 2015

on the use of T-spline local refinement in an analysis context Scott (2011); Scott et al. (2012); Borden et al. (2012); Verhoosel et al. (2011a,b).

T-splines, introduced in the CAD community (Sederberg et al. (2003)), are a generalization of non-uniform rational B-splines (NURBS) which address fundamental limitations in NURBS-based design. For example, a T-spline can model a complicated design as a single, watertight geometry and are also locally refineable (Sederberg et al. (2004); Scott et al. (2012)). Since their advent they have emerged as an important technology across multiple disciplines and can be found in several major commercial CAD products (Autodesk (2012); Autodesk, Inc. (2014)). Recent developments include analysis-suitable T-splines (Li et al. (2012); Scott et al. (2012); Beirão da Veiga et al. (2012); da Veiga et al. (2013); Li and Scott (2014)), and their hierarchical extension (Evans et al. (2014)).

In this chapter we focus on the design and analysis capabilities of T-splines. Commercial T-spline implementations exist for Rhino3d (Autodesk, Inc. (2011)) and several Autodesk products (Autodesk, Inc. (2014)). A T-spline plugin has been developed for Rhino3d (Autodesk, Inc. (2011)). The plugin is a useful tool for defining and exporting analysis-suitable T-spline models for use in isogeometric analysis. The primary capabilities include the definition of boundary condition sets based on vertex, edge, and face selection sets and export of analysis files for isogeometric finite element, boundary element, and collocation simulations. These files can be exported at any point during the design process and for any T-spline which can be defined in Rhino. No mesh generation or geometry cleanup is required. Additionally, the Bézier extraction of any T-spline as well as collocation point positions can be viewed through the Rhino interface.

2 T-splines

The following discussion of T-splines is taken from Thomas et al. (2014).

T-splines represent a generalization of B-splines and NURBS. Whereas a B-spline is constructed from a tensor product of univariate splines, a T-spline permits meshes with hanging nodes or T-junctions. T-splines contain standard B-splines and NURBS as special cases. The theory of T-splines is rich and dynamic. We do not delve deeply into T-spline theory in this chapter and instead refer the interested reader to Sederberg et al. (2004); Scott et al. (2011, 2012); Li and Scott (2014) and the references contained therein. We limit our discussion to two-dimensional T-splines of arbitrary degree.

2.1 The T-mesh

Given polynomial degrees, p_1 and p_2, and two index vectors, $\mathsf{I}_i = \{1, 2, \ldots, n_i + p_i + 1\}$, $i = 1, 2$, we define the index domain of the T-mesh as $\bar{\Omega} = [1, n_1 + p_1 + 1] \otimes [1, n_2 + p_2 + 1]$. We associate a knot vector $\mathsf{G}_i = \{\sigma_1^i \leq \sigma_2^i \leq \cdots \leq \sigma_{n_i+p_i+1}^i\}$, $i = 1, 2$, with the corresponding index vector I_i. Any repeated entries in G_i are referred to as knots of multiplicity m. We require that the first and last knots have multiplicity $p_i + 1$, this is commonly called an open knot vector. We also require that no knot in G_i have multiplicity greater than $p_i + 1$. It should be noted that repeated knots have unique indices. We define the parametric domain of the T-mesh as $\hat{\Omega} = [\sigma_1^1, \sigma_{n_1+p_1+1}^1] \otimes [\sigma_1^2, \sigma_{n_2+p_2+1}^2]$.

A T-mesh T is a rectangular partition of the index domain such that all vertices have integer coordinates taken from I_1 and I_2, all cells are rectangular, non-overlapping, and open, and all edges are horizontal or vertical line segments which do not intersect any cell. Because there are corresponding parametric values assigned to each vertex in the index space, the T-mesh can be mapped to the parametric domain. Cells in the index domain that are bounded by repeated knot values are mapped to cells of zero parametric area in the parametric domain. An example T-mesh is shown in fig. 1. Cells that are bounded by repeated knots in at least one dimension have zero parametric area and are shown in gray in the figure. We say that two

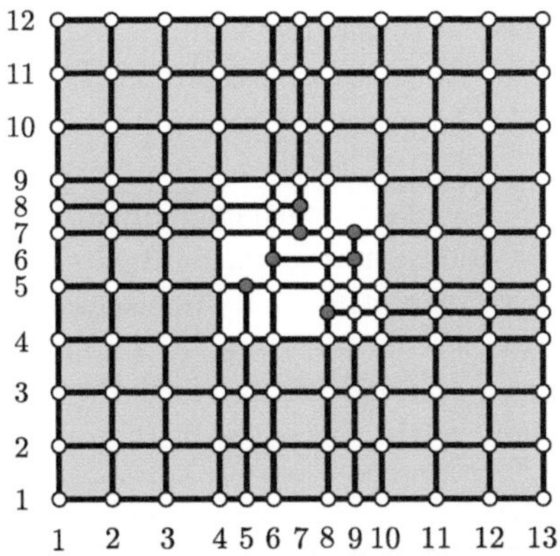

Figure 1: An example T-mesh. Vertices are marked with open circles and the vertices corresponding to T-junctions are marked with orange circles. Cells with zero parametric area are gray.

T-meshes T^a and T^b are nested if T^b can be created by adding vertices and edges to T^a. We use the notation $\mathsf{T}^a \subseteq \mathsf{T}^b$ to indicate this relationship.

2.2 T-spline basis functions

The T-spline basis functions are constructed from the T-mesh and the knot vectors. Note that we use the term basis function throughout this section although there is no guarantee that the set of blending functions inferred from a T-mesh form a basis for the space. This question is resolved by analysis-suitable T-splines. A basis function is anchored to unique T-mesh entities (i.e., vertices, edges, cells) as follows:

- If p_1 and p_2 are odd then the anchors are all the vertices in the T-mesh with indices greater than i_{p_1} and less than $n_1 + 1$ in the first dimension and greater than p_2 and less than $n_2 + 1$ in the second dimension.
- If p_1 and p_2 are even then the anchors are all the cells in the T-mesh bounded by vertices with indices greater than p_1 and less than $n_1 + 1$ in the first dimension and greater than p_2 and less than $n_2 + 1$ in the second dimension.
- If p_1 is even and p_2 is odd then the anchors are all the horizontal edges in the T-mesh bounded by vertices with indices greater than p_1 and less than $n_1 + 1$ in the first dimension and greater than p_2 and less than $n_2 + 1$ in the second dimension.
- If p_1 is odd and p_2 is even then the anchors are all the vertical edges in the T-mesh bounded by vertices with indices greater than p_1 and less than $n_1 + 1$ in the first dimension and greater than p_2 and less than $n_2 + 1$ in the second dimension.

The function anchors for these four cases are illustrated in fig. 2. We assume that the anchors can be enumerated and refer to each anchor by its index A.

The basis functions associated with each anchor are defined by constructing a local knot vector in each parametric direction. The function anchor is indicated by its index A and so the local knot vector associated with A in the ith parametric direction is denoted by $\mathbf{g}_{A,i}$. The algorithm for constructing the local knot vector in the ith parametric direction associated with the Ath anchor is given here with examples of its application following.

Algorithm 2.1. Construction of the local knot vector in the ith parametric direction for the Ath function anchor from the T-mesh.

1. Find the line that lies in the ith parametric direction and that passes through the center of the function anchor. We refer to this as the anchor line associated with the ith parametric direction. This line is used to find the indices that define the local knot vector.
2. Determine the width h_i of the anchor of interest in the direction perpendicular to the anchor line and thicken the line so that it has width

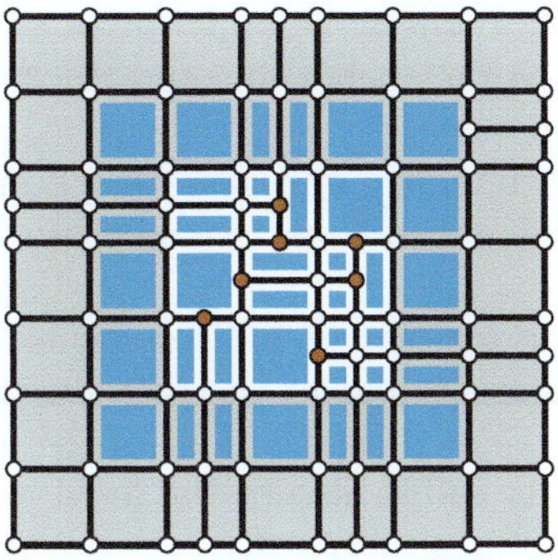

(a) $p_1 = 2$, $p_2 = 2$ anchors are faces.

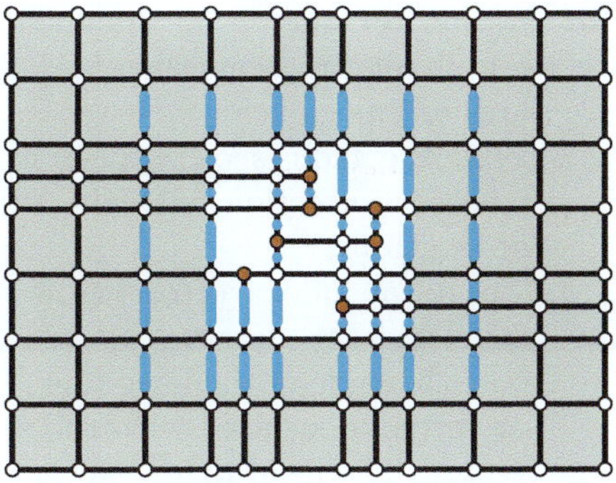

(b) $p_1 = 3$, $p_2 = 2$ anchors are vertical edges.

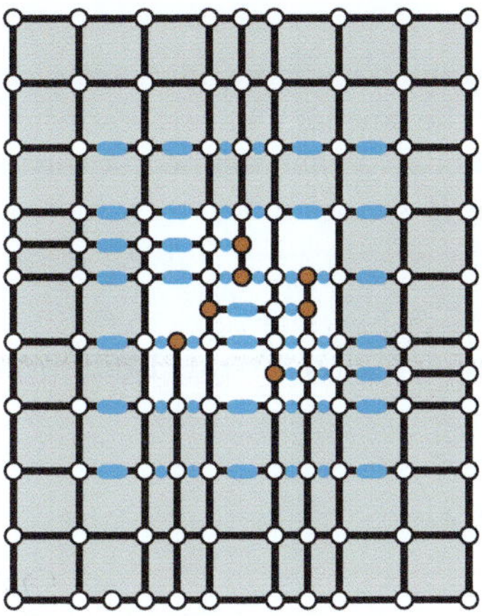

(c) $p_1 = 2$, $p_2 = 3$ anchors are horizontal edges.

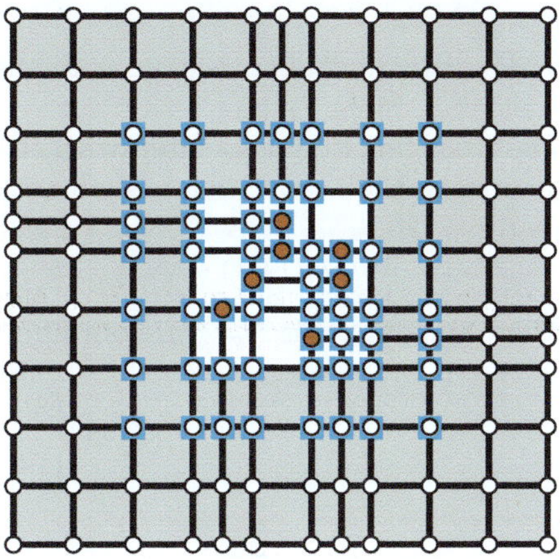

(d) $p_1 = 3$, $p_2 = 3$ anchors are vertices.

Figure 2: The set of anchors for varying values of p_1 and p_2. In this picture blue represents the anchor locations, gray is the boundary cells with zero parametric area, and the orange vertices are T-junctions.

h_i and is centered on the anchor line. If the polynomial degree p_i is odd in all directions then $h_i = 0$ and so the anchor line and thickened anchor line coincide.

3. Find the indices of vertices or perpendicular edges in the T-mesh whose intersection with the thickened anchor line is nonempty and of length h_i.
4. The local index vector $\mathsf{i}_{A,i}$ is the ordered set of indices formed by collecting the closest $\lceil (p_i+1)/2 \rceil$ indices found in the previous step on either side of the anchor A. If p_i is odd, then the index of the edge or vertex associated with the anchor is added also to the local index vector. The local index vector $\mathsf{i}_{A,i}$ is of length $p_i + 2$.
5. The local knot vector $\mathsf{g}_{A,i}$ is formed by collecting the knot entries in the global knot vector G_i given by the indices in the local index vector $\mathsf{i}_{A,i}$.

By carrying this process out in each parametric direction, a set of local knot vectors $\{\mathsf{g}_{A,1}, \mathsf{g}_{A,2}\}$ associated with the anchor A can be obtained. The basis function associated with the anchor A is then defined in the same manner as the local spline basis function for B-splines indexed by local knot vectors:

$$N_A(s,t) = N_{\mathsf{g}_{A,1},\mathsf{g}_{A,2}}(s,t) = N_{\mathsf{g}_{A,1}}(s)N_{\mathsf{g}_{A,2}}(t) \tag{1}$$

where the functions $N_{\mathsf{g}_{A,i}}$ are obtained by applying the Cox-de Boor formula

$$N_A^0(s) = \begin{cases} 1 & \sigma_A \leq s < \sigma_A \\ 0 & \text{otherwise.} \end{cases} \tag{2}$$

$$N_A^p(s) = \frac{s - \sigma_A}{\sigma_{A+p} - \sigma_A} N_A^{p-1}(s) + \frac{\sigma_{A+p+1} - s}{\sigma_{A+p+1} - \sigma_{A+1}} N_{A+1}^{p-1}(s) \tag{3}$$

where σ_A represents the Ath entry in the knot vector. The Cox-de Boor formula is applied to each of the local knot vectors $\mathsf{g}_{A,i}$ in turn. The process for constructing local knot vectors is illustrated in fig. 3 for four cases: $\vec{p} = \{2,2\}$, $\vec{p} = \{3,2\}$, $\vec{p} = \{2,3\}$, and $\vec{p} = \{3,3\}$.

T-splines with even degree in both directions have cell faces as anchors; thus for the $\vec{p} = \{2,2\}$ case shown in part (a), the function anchor is marked by a large, light blue box covering the cell face. We use the global knot vectors

$$\mathsf{G}_1 = \{0,0,0,1,2,3,4,5,6,7,7,7\} \tag{4}$$

and

$$\mathsf{G}_2 = \{0,0,0,1,2,3,4,5,6,7,7,7\}. \tag{5}$$

The anchor line used to construct the horizontal knot vector is shown in dark blue and the anchor line used to construct the vertical knot vector is shown in green. Because the spline has even polynomial degree, the thickened anchor line is shown in both directions. The indices that contribute to the local knot vector are marked with a $\times$. The indices used to construct the horizontal local index vector for the marked function are $\mathsf{i}_{A,1} = \{3, 4, 8, 10\}$ and so the local knot vector for the function is $\mathsf{g}_{A,1} = \{0, 1, 5, 7\}$. The indices 5 and 7 were skipped because there are no edges associated with those indices that intersect with the horizontal anchor line used to determine the local knot vector. Note that the index 9 was skipped because the edges that intersect the horizontal line do not span the thickened anchor line due to the missing edge between $(9, 6)$ and $(9, 7)$. Similarly, the vertical local index vector is $\mathsf{i}_{A,2} = \{2, 3, 7, 9\}$ and so the vertical local knot vector for the function is $\mathsf{g}_{A,2} = \{0, 0, 4, 6\}$.

The mixed degree case $\vec{p} = \{3, 2\}$ is shown in part (b). We now use the global knot vectors

$$\mathsf{G}_1 = \{0, 0, 0, 0, 1, 2, 3, 4, 5, 6, 7, 7, 7, 7\} \tag{6}$$

and

$$\mathsf{G}_2 = \{0, 0, 0, 1, 2, 3, 4, 5, 6, 7, 7, 7\}. \tag{7}$$

Here the function anchor is a vertical edge and so only the horizontal anchor line is thickened (shown in dark blue in the figure). The vertical anchor line is shown as a dashed green line. The horizontal indices that contribute to the local index vector are $\mathsf{i}_{A,1} = \{5, 9, 11, 12\}$ and so the horizontal local knot vector is $\mathsf{g}_{A,1} = \{1, 4, 5, 7, 7\}$. The index 7 is skipped because it does not have edges that span the thickened anchor line at the intersection. It is not necessary to check the span in the vertical direction because the anchor has no width. The vertical index vector is $\mathsf{i}_{A,2} = \{6, 7, 9, 10\}$ and the vertical local knot vector is $\mathsf{g}_{A,2} = \{3, 4, 6, 7\}$.

The opposite mixed degree case $\vec{p} = \{2, 3\}$ is shown in part (c). The global knot vectors are now

$$\mathsf{G}_1 = \{0, 0, 0, 1, 2, 3, 4, 5, 6, 7, 7, 7\} \tag{8}$$

and

$$\mathsf{G}_2 = \{0, 0, 0, 0, 1, 2, 3, 4, 5, 6, 7, 7, 7, 7\}. \tag{9}$$

The function anchors are now horizontal edges and so only the vertical anchor line must be thickened. The indices for the horizontal local index vector are $\mathsf{i}_{A,1} = \{3, 4, 7, 8\}$ and the knot vector is $\mathsf{g}_{A,1} = \{0, 1, 4, 5\}$. The horizontal anchor line has no width perpendicular to the horizontal direction

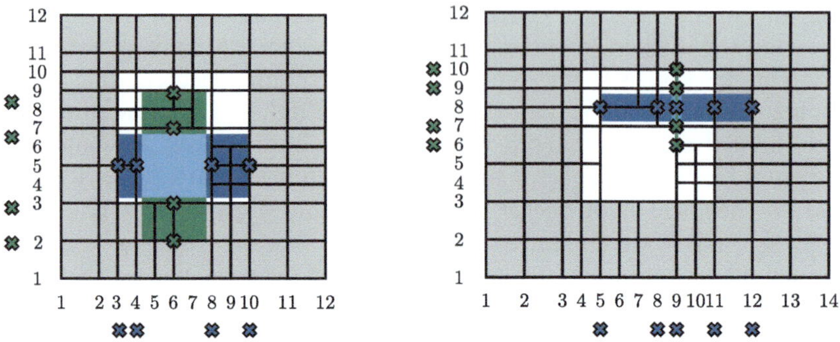

(a) $p_1 = 2$, $p_2 = 2$ anchors are faces. Thickened anchor lines in both directions.

(b) $p_1 = 3$, $p_2 = 2$ anchors are vertical edges. Only the horizontal anchor line must be thickened.

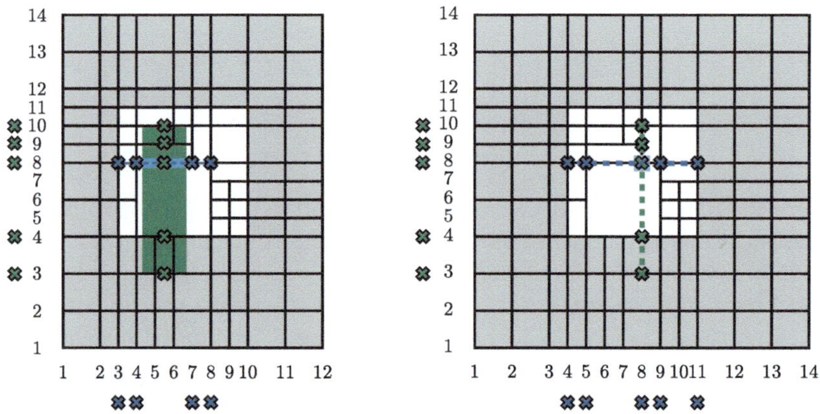

(c) $p_1 = 2$, $p_2 = 3$ anchors are horizontal edges. Only the vertical anchor line must be thickened.

(d) $p_1 = 3$, $p_2 = 3$ anchors are vertices. Neither anchor line must be thickened.

Figure 3: Examples for how local knot vectors are constructed for T-spline basis functions of varying values of the polynomial degrees p_1 and p_2. The function anchors are marked with light blue. The thickened horizontal anchor line used to determine the horizontal knot vector is indicated with a dark box where necessary while in cases that do not require thickening it is shown as a dark blue dashed line. The vertical anchor line used to calculate the vertical knot vector is marked in green. The indices that contribute to the local knot vectors are marked with $\times$ and colored dark blue for those in the horizontal direction and green for those in the vertical direction.

and so only intersections must be checked. The indices for the vertical local index vector are $\mathsf{i}_{A,2} = \{3, 4, 8, 9, 10\}$; here all of the edges intersected span the anchor. The vertical local knot vector is $\mathsf{g}_{A,2} = \{0, 0, 4, 5, 6\}$.

The odd degree case $\vec{p} = \{3, 3\}$ is shown in part (d). We choose the global knot vectors

$$\mathsf{G}_1 = \{0, 0, 0, 0, 1, 2, 3, 4, 5, 6, 7, 7, 7, 7\} \tag{10}$$

and

$$\mathsf{G}_2 = \{0, 0, 0, 0, 1, 2, 3, 4, 5, 6, 7, 7, 7, 7\}. \tag{11}$$

For T-splines with odd degree in both directions, the anchors are vertices, the anchor lines are not thickened, and so only intersection must be checked. The indices for the horizontal local knot vector are $\{4, 5, 8, 9, 11\}$ and the local knot vector is $\mathsf{g}_{A,1} = \{0, 1, 4, 5, 7\}$. The indices for the vertical local knot vector are $\{3, 4, 8, 9, 10\}$ and so the local knot vector is $\mathsf{g}_{A,2} = \{0, 0, 4, 5, 6\}$. Once the T-spline basis functions have been defined and control points have been assigned to each one, the geometric map is given by

$$\vec{x}(s, t) = \sum_{A=1}^{n} \vec{P}_A N_A(s, t). \tag{12}$$

2.3 Face and edge extensions and analysis-suitable T-splines

Although the T-spline blending functions defined in this fashion can be used to define smooth surfaces, many properties of the resulting space are not immediately obvious. In order to develop T-splines that are well-characterized and suitable for analysis, we introduce the face and edge extensions of T-junctions. A face extension is a closed line segment that extends from the T-junction in the direction of the face of the cell at which the T-junction terminates and that crosses $\lfloor (p_i + 1)/2 \rfloor$ perpendicular edges or vertices. The edge extension of a T-junction is a closed line segment that extends from the T-junction in the opposite direction of the face extension and that crosses $\lceil (p_i - 1)/2 \rceil$ perpendicular edges or vertices. A T-mesh is analysis suitable if no vertical T-junction extension intersects a horizontal T-junction extension. Note that because the edge and face extensions are closed segments, they can intersect at endpoints. The face and edge extensions are shown for a T-mesh that is not analysis-suitable and for an analysis-suitable T-mesh in fig. 4. The T-mesh formed by adding all face extensions to T is called the extended T-mesh and is denoted by T_{ext}.

The T-spline basis defined by an analysis-suitable T-mesh possesses many important mathematical properties including the following theorems (Evans et al. (2014); Li and Scott (2014)):

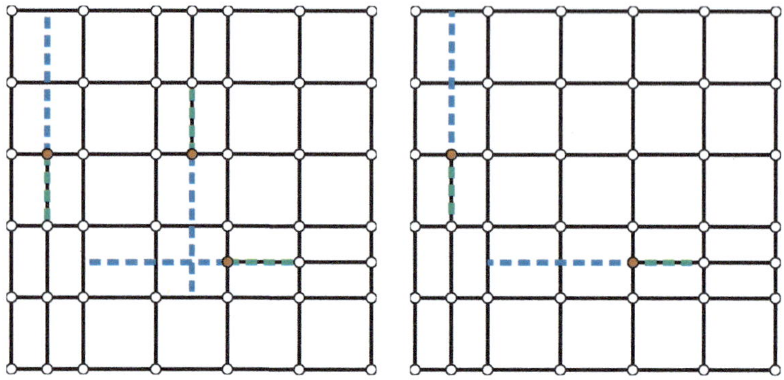

Figure 4: T-junction extension in two dimensions for a bicubic T-spline. Face extensions are shown in blue and edge extensions are shown in green. The T-junctions are marked with orange circles. The T-mesh on the left is not analysis-suitable while the T-mesh on the right is.

Theorem 2.2. *The basis functions of an analysis-suitable T-spline are locally linearly independent.*

Theorem 2.3. *The basis functions of an analysis-suitable T-spline form a complete basis for the space of polynomials of degree $\vec{p}$.*

Theorem 2.4. *The analysis-suitable T-spline spaces $\mathcal{T}^a$ and $\mathcal{T}^b$ are nested (i.e., $\mathcal{T}^a \subseteq \mathcal{T}^b$) if $\mathsf{T}^a_{\text{ext}} \subseteq \mathsf{T}^b_{\text{ext}}$, that is, if $\mathsf{T}^b_{\text{ext}}$ can be constructed by adding edges to $\mathsf{T}^a_{\text{ext}}$.*

We only consider analysis-suitable T-splines (ASTS) for the remainder of this work. The spline space spanned by a T-spline basis defined by the T-mesh T^a is denoted by $\mathcal{T}^a$.

This concludes the discussion of T-splines taken from Thomas et al. (2014).

2.4 Bézier extraction

In this chapter we develop T-splines from the finite element point-of-view, utilizing Bézier extraction. Bézier extraction for T-splines was described in detail in (Scott et al. (2011)). The idea is to extract the linear operator which maps the Bernstein polynomial basis on Bézier elements to the global T-spline basis. The linear transformation is defined by a matrix referred to as the extraction operator and denoted by $\mathbf{C}^e$. The transpose of the extraction operator maps the control points of the global T-spline to the control points of the Bernstein polynomials. Figure 5 illustrates the idea for a B-spline curve. This provides a finite element representation of T-splines, and facilitates the incorporation of T-splines into existing finite element programs. Only the shape function subroutine needs to be modified. All other aspects of the finite element program remain the same. Additionally, Bézier extraction is automatic and can be applied to any T-spline regardless of topological complexity or polynomial degree. In particular, it represents an elegant treatment of T-junctions, referred to as "hanging nodes" in finite element analysis.

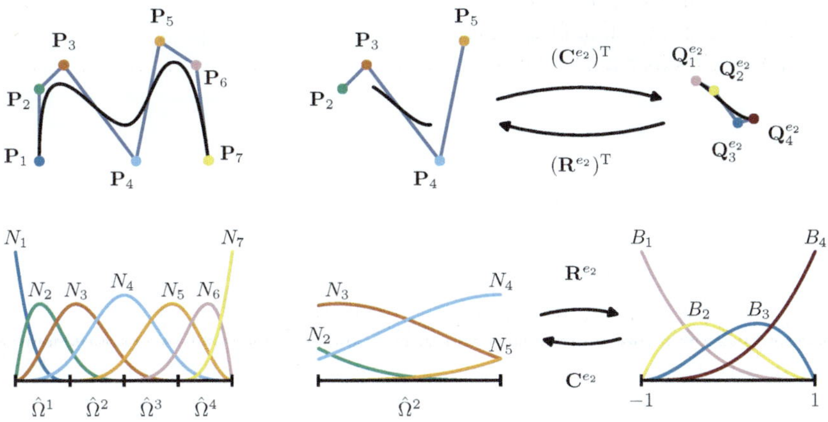

Figure 5: Schematic representation of Bézier extraction for a B-spline curve. B-spline basis functions and control points are denoted by **N** and **P**, respectively. Bernstein polynomials and control points are denoted by **B** and **Q**, respectively. The curve $T(\boldsymbol{\xi}) = \mathbf{P}^T\mathbf{N}(\boldsymbol{\xi}) = \mathbf{Q}^T\mathbf{B}(\boldsymbol{\xi})$.

2.5 Representation of a T-spline basis by Bézier extraction

As explained previously, a T-spline basis function, N_A, is defined for every vertex, A, in the T-mesh. Each N_A is a bivariate piecewise polynomial function. Several bicubic T-spline basis functions are shown in Figure 6. T-spline basis functions are often represented in Bézier form. Bézier extraction can be applied to an entire T-spline to generate a finite set of Bézier elements such that

$$\mathbf{N}^e(\boldsymbol{\xi}) = \mathbf{C}^e\mathbf{B}(\boldsymbol{\xi}), \tag{13}$$

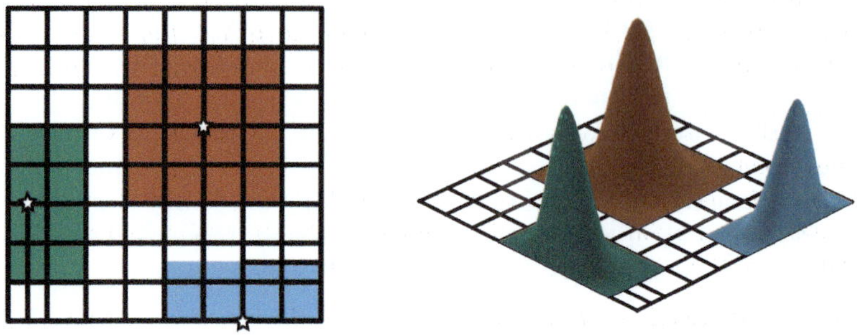

Figure 6: Bicubic T-spline basis functions

where $\boldsymbol{\xi} = (\xi, \eta, \zeta) \in \tilde{\Omega}$ is a coordinate in a standard Bézier parent element domain (see Hughes (2000), Chapter 3, for a review of standard finite element paraphernalia, including the parent element domain), $\mathbf{N}^e(\boldsymbol{\xi}) = \{N_a^e(\boldsymbol{\xi})\}_{a=1}^n$ is a vector of T-spline basis functions which are non-zero over Bézier element e, $\mathbf{B}(\boldsymbol{\xi}) = \{B_i(\boldsymbol{\xi})\}_{i=1}^m$ is a vector of tensor product Bernstein polynomial basis functions defining Bézier element e, and $\mathbf{C}^e \in \mathbb{R}^{n \times m}$ is the element extraction operator.

2.6 The T-spline geometric map

We can define the element geometric map, $\mathbf{x}^e : \tilde{\Omega} \to \Omega^e$, from the parent element domain onto the physical domain as

$$\mathbf{x}^e(\boldsymbol{\xi}) = \frac{1}{(\mathbf{w}^e)^T \mathbf{N}^e(\boldsymbol{\xi})} (\mathbf{P}^e)^T \mathbf{W}^e \mathbf{N}^e(\boldsymbol{\xi}) \tag{14}$$

$$= (\mathbf{P}^e)^T \mathbf{R}^e(\boldsymbol{\xi}) \tag{15}$$

where $\mathbf{R}^e(\boldsymbol{\xi}) = \{R_a^e(\boldsymbol{\xi})\}_{a=1}^n$ is a vector of rational T-spline basis functions, the element weight vector $\mathbf{w}^e = \{w_a^e\}_{a=1}^n$, the diagonal weight matrix $\mathbf{W}^e = diag(\mathbf{w}^e)$, and $\mathbf{P}^e$ is a matrix of dimension $n \times d_s$ that contains element control points,

$$\mathbf{P}^e = \begin{bmatrix} P_1^{e,1} & P_1^{e,2} & \dots & P_1^{e,d_s} \\ P_2^{e,1} & P_2^{e,2} & \dots & P_2^{e,d_s} \\ \vdots & \vdots & & \vdots \\ P_n^{e,1} & P_n^{e,2} & \dots & P_n^{e,d_s} \end{bmatrix}. \tag{16}$$

Using (14) and (15) we have that

$$\mathbf{R}^e(\boldsymbol{\xi}) = \frac{1}{(\mathbf{w}^e)^T \mathbf{N}^e(\boldsymbol{\xi})} \mathbf{W}^e \mathbf{N}^e(\boldsymbol{\xi}), \tag{17}$$

and using (13)

$$\mathbf{R}^e(\boldsymbol{\xi}) = \frac{1}{(\mathbf{w}^e)^T \mathbf{C}^e \mathbf{B}(\boldsymbol{\xi})} \mathbf{W}^e \mathbf{C}^e \mathbf{B}(\boldsymbol{\xi}). \tag{18}$$

Note that all quantities in (18) are written in terms of the Bernstein basis defined over the parent element domain, $\tilde{\Omega}$.

2.7 The Bézier mesh

The Bézier control points $\mathbf{Q}^e$ and Bézier weights $\mathbf{w}^{b,e}$ for element e are computed as

$$\mathbf{Q}^e = \left(\mathbf{W}^{b,e}\right)^{-1} (\mathbf{C}^e)^T \mathbf{W}^e \mathbf{P}^e, \tag{19}$$

$$\mathbf{w}^{b,e} = (\mathbf{C}^e)^T \mathbf{w}^e, \tag{20}$$

where $\mathbf{P}^e$ are the T-spline control points, $\mathbf{W}^{b,e}$ is the diagonal matrix of Bézier weights, and $\mathbf{W}^e$ is the diagonal matrix of T-spline weights, $\mathbf{w}^e$. Figure 7 shows the result of (19) for several Bézier elements in a T-spline. The T-spline control points which contribute to the location of the Bézier element control points are indicated by the $\bigcirc$'s. Each T-spline element has an equivalent representation as an extracted Bézier element. In other words

$$\mathbf{x}^e(\boldsymbol{\xi}) = \mathbf{x}^{b,e}(\boldsymbol{\xi}) = \frac{1}{W^{b,e}(\boldsymbol{\xi})} (\mathbf{Q}^e)^T \mathbf{W}^{b,e} \mathbf{B}(\boldsymbol{\xi}). \tag{21}$$

where $W^{b,e}(\boldsymbol{\xi}) = \mathbf{w}^{b,e\,T} \mathbf{B}(\boldsymbol{\xi})$.

3 Integrating Bézier extraction and isogeometric analysis

Bézier extraction of T-splines generates a set of Bézier elements (written in terms of the Bernstein basis) and the corresponding element extraction

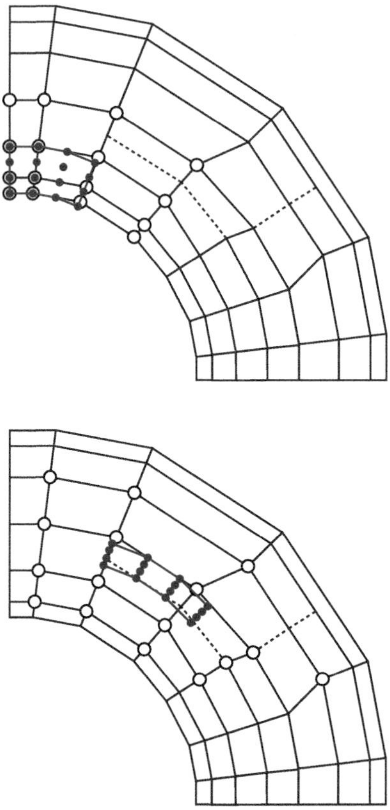

Figure 7: For an element e the ○'s indicate the global T-spline control points which influence the location of Bézier control points, indicated by the •'s. The global control points that influence each element are determined through Bézier extraction and the location of Bézier control points is computed with the element extraction operator $\mathbf{C}^e$.

operators, $\mathbf{C}^e$, which can be integrated into a finite element framework in a straightforward manner. Starting with an abstract weak formulation,

$$(W)\begin{cases} \text{Given } f, \text{ find } u \in \mathcal{S} \text{ such that for all } w \in \mathcal{V} \\ \qquad a(w,u) = (w,f) \end{cases} \tag{22}$$

where $a(\cdot,\cdot)$ is a bilinear form and $(\cdot,\cdot)$ is the L^2 inner-product and $\mathcal{S}$ and $\mathcal{V}$ are the trial solution space and space of weighting functions, respectively, Galerkin's method consists of constructing finite-dimensional approximations of $\mathcal{S}$ and $\mathcal{V}$. In isogeometric analysis these finite-dimensional subspaces $\mathcal{S}^h \subset \mathcal{S}$ and $\mathcal{V}^h \subset \mathcal{V}$ are constructed from the T-spline basis which describes the geometry. The Galerkin formulation is then

$$(G)\begin{cases} \text{Given } f, \text{ find } u^h \in \mathcal{S}^h \text{ such that for all } w^h \in \mathcal{V}^h \\ \qquad a(w^h,u^h) = (w^h,f) \end{cases} \tag{23}$$

In isogeometric analysis, the isoparametric concept is invoked, that is, the field in question is represented in terms of the geometric T-spline basis. We can write u^h and w^h as

$$w^h = \sum_{A=1}^{n} c_A R_A \tag{24}$$

$$u^h = \sum_{B=1}^{n} d_B R_B \tag{25}$$

where c_A and d_B are control variables. Substituting these into (23) yields the matrix form of the problem

$$\mathbf{Kd} = \mathbf{F} \tag{26}$$

where

$$\mathbf{K} = [K_{AB}], \tag{27}$$

$$\mathbf{F} = \{F_A\}, \tag{28}$$

$$\mathbf{d} = \{d_B\}, \tag{29}$$

$$K_{AB} = a\left(R_A, R_B\right), \tag{30}$$

$$F_A = \left(R_A, f\right). \tag{31}$$

The preceding formulation applies to scalar-valued partial differential equations, such as the heat conduction equation. The generalization to vector-valued partial differential equations, such as elasticity, follows standard procedures as described in Hughes (2000).

As in standard finite elements, the global stiffness matrix, $\mathbf{K}$, and force vector, $\mathbf{F}$, can be constructed by performing integration over the Bézier elements to form element stiffness matrices and force vectors, $\mathbf{k}^e$ and $\mathbf{f}^e$, respectively, and assembling these into their global counterparts. The element form of (30) and (31) is

$$k^e_{ab} = a_e(R^e_a, R^e_b), \tag{32}$$

$$f^e_a = (R^e_a, f)_e \tag{33}$$

where $a_e(\cdot,\cdot)$ denotes the bilinear form restricted to the element, $(\cdot,\cdot)_e$ is the L^2 inner-product restricted to the element, and R^e_a are the element shape functions. The integration is usually performed by Gaussian quadrature. Since T-splines are not, in general, defined over a global parametric domain, all integrals are pulled back directly to the bi-unit parent element domain. No intermediate parametric domain is employed. This requires the evaluation of the global T-spline basis functions, their derivatives, and the Jacobian determinant of the pullback from the physical space to the parent element at each quadrature point in the parent element. These evaluations are done in an element shape function routine. Employing the element extraction operators we have that,

$$\mathbf{R}^e(\boldsymbol{\xi}) = \mathbf{W}^e\mathbf{C}^e\frac{\mathbf{B}(\boldsymbol{\xi})}{W^e(\boldsymbol{\xi})}. \tag{34}$$

where

$$W^e(\boldsymbol{\xi}) = (\mathbf{w}^e)^T\mathbf{C}^e\mathbf{B}(\boldsymbol{\xi}). \tag{35}$$

The derivatives of $\mathbf{R}^e$ with respect to the local parent coordinates, ξ^i, are

$$\frac{\partial\mathbf{R}^e(\boldsymbol{\xi})}{\partial\xi^i} = \mathbf{W}^e\mathbf{C}^e\frac{\partial}{\partial\xi^i}\left(\frac{\mathbf{B}(\boldsymbol{\xi})}{W^e(\boldsymbol{\xi})}\right) = \mathbf{W}^e\mathbf{C}^e\left(\frac{1}{W^e(\boldsymbol{\xi})}\frac{\partial\mathbf{B}(\boldsymbol{\xi})}{\partial\xi^i} - \frac{\partial W^e(\boldsymbol{\xi})}{\partial\xi^i}\frac{\mathbf{B}(\boldsymbol{\xi})}{(W^e(\boldsymbol{\xi}))^2}\right). \tag{36}$$

To compute the derivatives with respect to the physical coordinates, $(\tilde{x}^e_1, \tilde{x}^e_2, \tilde{x}^e_3)$, we apply the chain rule to get

$$\frac{\partial\mathbf{R}^e(\boldsymbol{\xi})}{\partial\tilde{x}^e_i} = \sum_{j=1}^{3}\frac{\partial\mathbf{R}^e(\boldsymbol{\xi})}{\partial\xi^j}\frac{\partial\xi^j}{\partial\tilde{x}^e_i}. \tag{37}$$

To compute $\partial \boldsymbol{\xi}/\partial \tilde{\mathbf{x}}^e$ we first compute $\partial \tilde{\mathbf{x}}^e/\partial \boldsymbol{\xi}$ using (15) and (36) and then take its inverse. Since we are integrating over the parent element we must also compute the Jacobian determinant, J^e, of the mapping from the parent element to the physical space. It is computed as

$$J^e = \left| \frac{\partial \tilde{\mathbf{x}}^e}{\partial \boldsymbol{\xi}} \right| . \tag{38}$$

Higher-order derivatives can also be computed as described in Cottrell et al. (2009).

4 Applications in CAD and analysis

This section provides a discussion of two examples of the application of IGA using T-splines in analysis. The majority of the first section is quoted from Scott et al. (2013) and the interested reader is referred to that paper for more details. The second section is quoted from Scott (2011) and contains material also presented by Benson et al. (2010a).

4.1 Analysis of a propeller

We now consider the application of IGA to the analysis of a propeller design. An isogeometric boundary element method (IGA BEM) is employed. This permits analysis to be conducted using a surface representation. IGA BEM uses native T-spline surface descriptions directly circumventing traditional mesh generation procedures

In order to represent complex geometries it is necessary to introduce extraordinary vertices, also known as extraordinary points. The valence of a vertex, denoted by N, is the number of edges that touch the vertex. An extraordinary vertex is an interior vertex that is not a T-junction and for which $N \neq 4$. Figure 8 shows an example T-mesh. The T-junction is denoted by a red square while the extraordinary vertices are denoted by red circles.

To illustrate the ability of the isogeometric boundary element method to handle complex engineering geometries, a T-spline propeller design is analyzed. The extracted Bézier mesh is shown in Figure 9(a). The entire geometry is a single T-spline consisting of $10,080$ control points and $10,080$ Bézier elements. Figure 9(b) shows a closeup of an unstructured section of the Bézier mesh. Notice the smooth Bézier elements surrounding each extraordinary point. The construction of these elements is beyond the scope of this paper. We refer the interested reader to the detailed description in Scott et al. (2013). There are a total of 48 extraordinary points in the model.

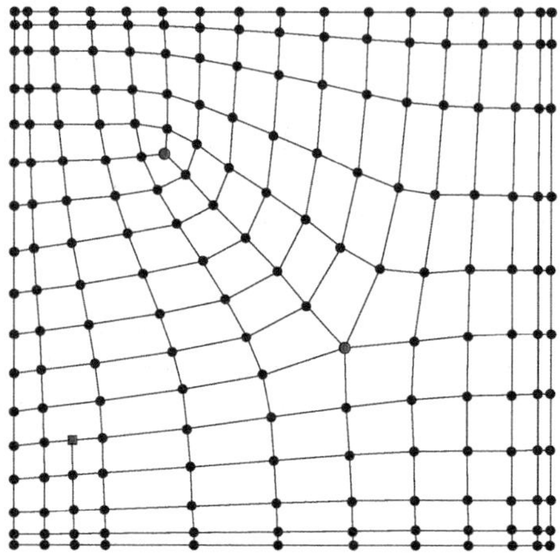

Figure 8: An example T-mesh. Extraordinary points are denoted by red hollow circles and T-junctions are denoted by red hollow squares.

To illustrate the importance of being able to accommodate extraordinary points in both design and analysis, an equivalent representation of the propeller as a multi-patch NURBS is shown in Figure 9(c). In this case, there are 72 NURBS patches. In the context of NURBS-based design, the propeller would be constructed patch-by-patch with smoothness across patches enforced (usually manual) by control point positioning. When used in analysis, all inter-patch connectivity and smoothness must be enforced explicitly in the analysis to ensure consistent (at least continuous) deformations of the geometry. This would be an enormous burden in analysis, but is completely eliminated with T-splines.

In the context of T-spline-based isogeometric analysis, extraordinary points are handled naturally and all smoothness considerations are built into the basis functions. These properties are transferred automatically to the analysis since the geometric basis is used as the basis for analysis.

We note that the multi-patch NURBS shown in Figure 9(c) was generated automatically from the T-spline in Figure 9(a). This forward and backward compatibility between T-splines and NURBS is made possible, in large part, by the finite polynomial representation of the irregular Bézier elements surrounding each extraordinary point as described in Section 2.5. This would not be possible if a subdivision-based approach was being used near the extraordinary points.

A wind loading is simulated by setting all displacement components on

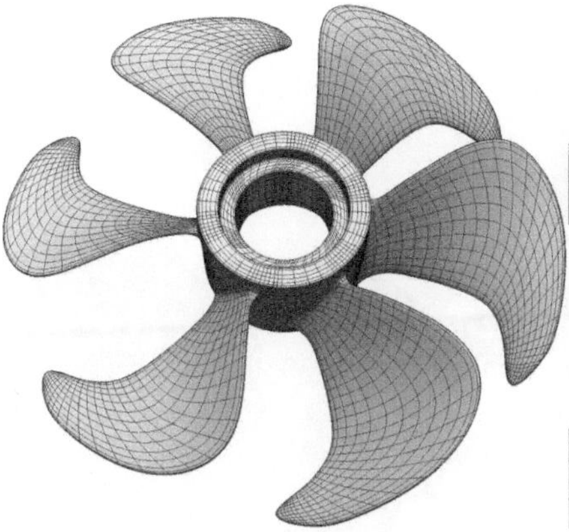

(a) An extracted Bézier mesh for a propeller.

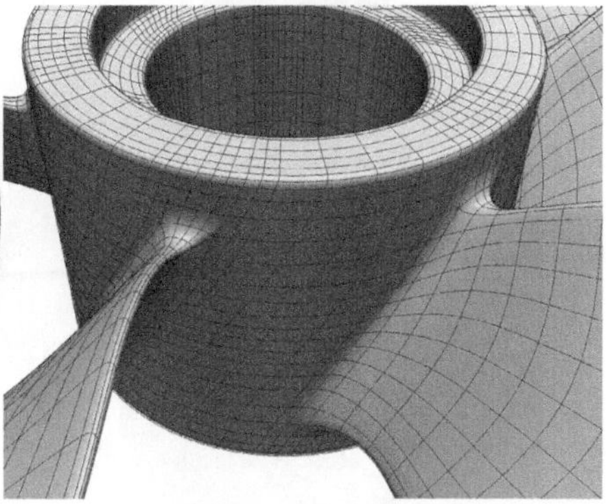

(b) A closeup of a smooth unstructured region of the Bézier mesh. The transition between each propeller fin and the shaft is accomplished using four extraordinary points.

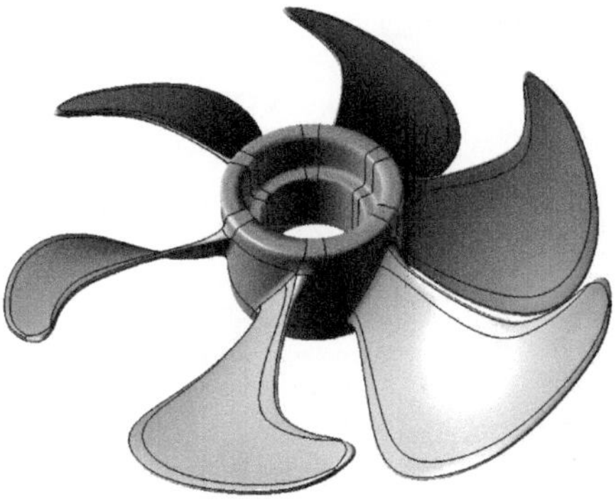

(c) An equivalent 72 patch NURBS for the T-spline propeller in Figure 9(a).

Figure 9: Geometries used in isogeometric analysis of a propeller.

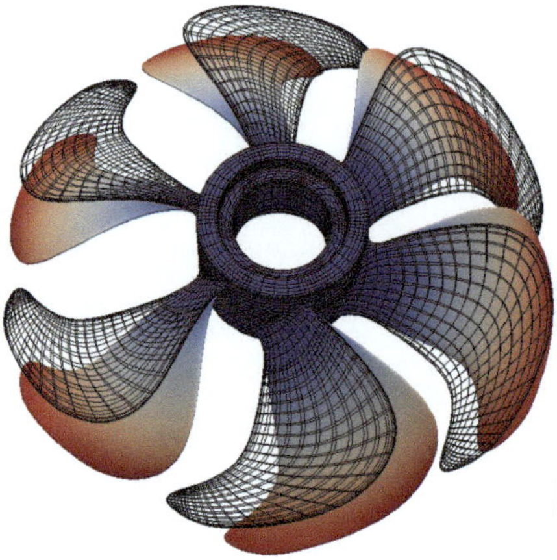

(a) The original Bézier mesh for the propeller model and an exaggerated displacement profile with displacement magnitude superimposed.

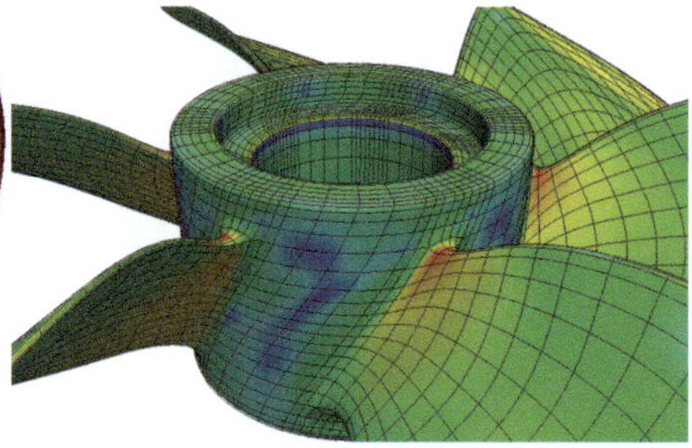

(b) The von Mises stress solution near extraordinary points

(c) The von Mises stress solution for the T-spline propeller model. Note that a logarithmic scale is used to plot the stress

Figure 10: Displacement and stress predicted by isogeometric analysis of a propeller.

the interior cylindrical surface to zero and setting the traction on all other surfaces to $\mathbf{t} = \{0, 0, -1000n_z\}^T$ if $n_z > 0$ and zero otherwise. We set Young's modulus, E, to 1e5 and Poisson's ratio, ν, to 0.3. Figure 10(a) shows the undeformed configuration and exaggerated displacement profile with the magnitude of the resulting displacement field superimposed. The von Mises stress profile is shown in Figure 10(c). Note that a logarithmic scale for the stress is used to more clearly visualize the changes in stress over the surface. A closeup of the von Mises stress solution near extraordinary points is shown in Figure 10(b). Notice the smoothness of the resulting stress profile.

4.2 Design and analysis integration

An example which effectively synthesizes the basic ideas and demonstrates, on a high level, the potential of T-spline-based isogeometric analysis as an integrative technology is shown in Figures 11 and 12. The car bumper model was generated using the T-spline plug-in for Rhino (Autodesk, Inc. (2011)) and the back-end T-Tools libraries (T-Splines, Inc. (2011)). In Figure 11a a bi-cubic NURBS is used to model the general shape of the bumper. To add features, such as holes, it is common to use trimming curves. Figures 11b and 11c show the placement of a trimming curve and the resulting geometry once the trim is applied. In a design-through-analysis (DTA) environment, the application of trimming curves destroys the analysis-suitable nature of the geometry. The geometric basis no longer describes the geometry and cannot be used directly in analysis.

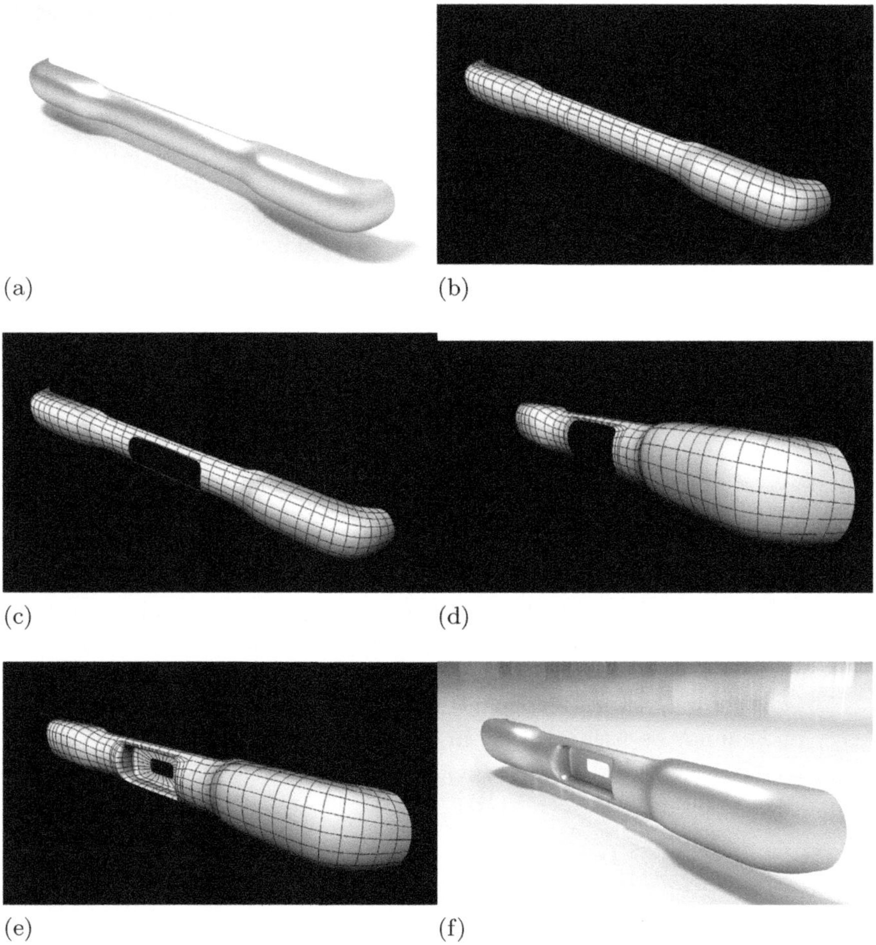

Figure 11: The design of an analysis-suitable T-spline bumper model. Bumper model courtesy of Juan Santocono.

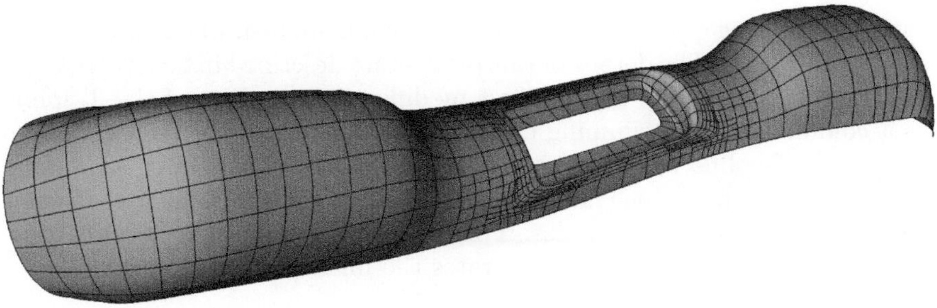

(a) Mode 7

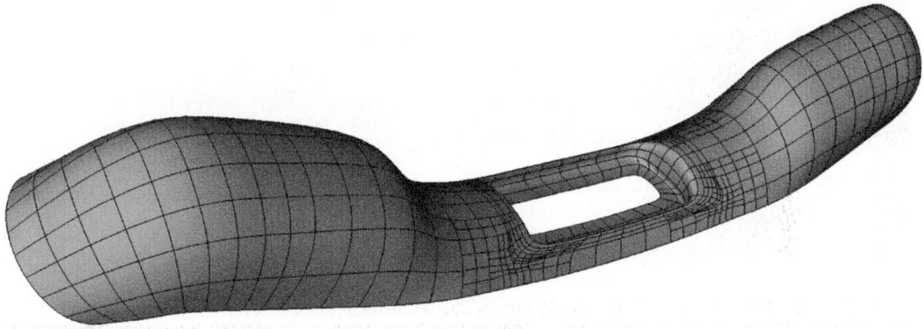

(b) Mode 9

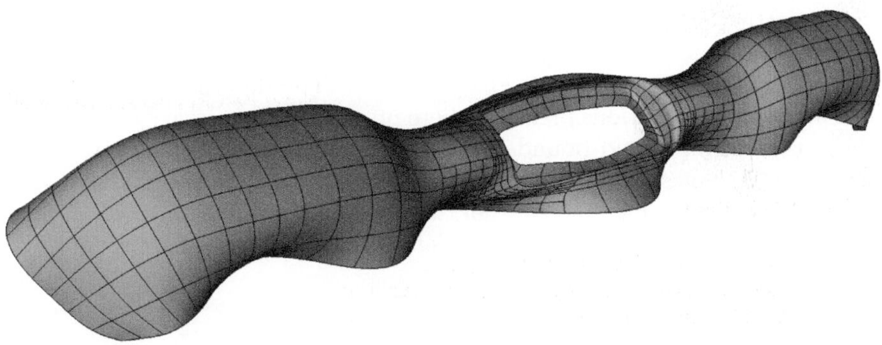

(c) Mode 30

Figure 12: Modes 7, 9, and 30 for the bumper model.

Using T-splines, however, it is possible to overcome the trimming problem by converting a trimmed T-spline into an untrimmed, watertight, analysis-suitable T-spline. The details of this process are described in Sederberg et al. (2008). The conversion process first modifies the topology of the T-spline to accommodate any trimming curves. A fitting procedure is then used to match the T-spline surface to the trimming curve. Figure 11d shows the untrimmed T-spline which matches the original trimmed T-spline upon completion of the conversion process. This untrimmed T-spline is now analysis-suitable. Additional modeling generates the final bumper geometry shown in Figures 11e and 11f.

The final model of the bumper consists of 876 cubic T-spline shell elements with 705 control points. No intermediate geometry clean-up or meshing step was employed. The free-free eigenvalues were calculated. The calculations were performed directly on the T-spline model in LS-DYNA. See Benson et al. (2010b,a) for further details. The seventh, ninth and thirtieth modes are shown in Figure 12.

5 Conclusion

Isogeometric analysis based on T-splines provides a strong foundation for integrated design through analysis. The existence of commercial-grade tools that are capable of exporting analysis-ready geometry representations eliminates the need for meshing in many cases. We have described the definition and construction of T-spline basis functions and analysis-suitable T-splines. The use of Bézier extraction permits T-spline technology to be integrated into traditional element-based analysis tools. The description of T-spline bases and shape functions presented here provides the necessary background to implement finite and boundary element analysis tools that leverage the power of T-splines. Two examples of integrated design and analysis were given: boundary element analysis of the stress in a complex propeller design represented as a T-spline and finite element shell analysis of a bumper.

Bibliography

Autodesk. *Autodesk T-Splines Plug-in for Rhino user manual.* Autodesk, 2012.

Autodesk, Inc. http://www.tsplines.com/rhino/, 2011.

Autodesk, Inc. *Autodesk Fusion 360.* Autodesk, Inc., 2014.

Y. Bazilevs, V. M. Calo, J. A. Cottrell, J. A. Evans, T. J. R. Hughes, S. Lipton, M. A. Scott, and T. W. Sederberg. Isogeometric analysis using

T-splines. *Computer Methods in Applied Mechanics and Engineering*, 199 (5-8):229–263, 2010. ISSN 0045-7825.

Y. Bazilevs, M. C. Hsu, and M. A. Scott. Isogeometric fluid-structure interaction analysis with emphasis on non-matching discretizations, and with application to wind turbines. *Computer Methods in Applied Mechanics and Engineering*, 249 - 252:28 – 41, 2012.

L. Beirão da Veiga, A. Buffa, D. Cho, and G. Sangalli. Analysis-suitable T-splines are dual-compatible. *Computer Methods in Applied Mechanics and Engineering*, 249 – 252:42 – 51, 2012.

D. J. Benson, Y. Bazilevs, E. De Luycker, M. C. Hsu, M. A. Scott, T. J. R Hughes, and T. Belytschko. A generalized finite element formulation for arbitrary basis functions: From isogeometric analysis to XFEM. *International Journal for Numerical Methods in Engineering*, 83:765–785, 2010a.

D. J. Benson, Y. Bazilevs, M. C. Hsu, and T. J. R. Hughes. Isogeometric shell analysis: The Reissner-Mindlin shell. *Computer Methods in Applied Mechanics and Engineering*, 199(5-8):276–289, 2010b.

M. J. Borden, M. A. Scott, C. V. Verhoosel, C. M. Landis, and T. J. R. Hughes. A phase-field description of dynamic brittle fracture. *Computer Methods in Applied Mechanics and Engineering*, 217:77 – 95, 2012.

A. Buffa, G. Sangalli, and R. Vázquez. Isogeometric methods for computational electromagnetics: B-spline and T-spline discretizations. *Journal of Computational Physics*, 257, Part B:1291 – 1320, 2014.

E. Cohen, T. Martin, R. M. Kirby, T. Lyche, and R. F. Riesenfeld. Analysis-aware modeling: Understanding quality considerations in modeling for isogeometric analysis. *Computer Methods in Applied Mechanics and Engineering*, 199(5-8):334–356, 2010.

J. A. Cottrell, T. J. R. Hughes, and A. Reali. Studies of refinement and continuity in isogeometric analysis. *Computer Methods in Applied Mechanics and Engineering*, 196:4160–4183, 2007.

J. A. Cottrell, T. J. R. Hughes, and Y. Bazilevs. *Isogeometric analysis: Toward Integration of CAD and FEA*. Wiley, Chichester, 2009.

L. B. da Veiga, A. Buffa, G. Sangalli, and R. Vázquez. Analysis-suitable T-splines of arbitrary degree: definition, linear independence, and approximation properties. *Mathematical Models and Methods in Applied Sciences*, 23(11):1979 – 2003, 2013.

R. Dimitri, L. De Lorenzis, M. A. Scott, P. Wriggers, R. Taylor, and G. Zavarise. Isogeometric large deformation frictionless contact using T-splines. *Computer methods in applied mechanics and engineering*, 269: 394 – 414, 2014.

M. Dörfel, B. Jüttler, and B. Simeon. Adaptive isogeometric analysis by local h-refinement with T-splines. *Computer Methods in Applied Mechanics and Engineering*, 199(5–8):264–275, 2009.

E. J. Evans, M. A. Scott, X. Li, and D. C. Thomas. Hierarchical analysis-suitable T-splines: Formulation, Bézier extraction, and application as an adaptive basis for isogeometric analysis. *submitted*, 2014.

J. A. Evans, Y. Bazilevs, I. Babuška, and T. J. R. Hughes. n-widths, sup-infs, and optimality ratios for the k-version of the isogeometric finite element method. *Computer Methods in Applied Mechanics and Engineering*, 198(21-26):1726–1741, 2009.

A. I. Ginnis, K. V. Kostas, C. G. Politis, P. D. Kaklis, K. A. Belibassakis, Th. P. Gerostathis, M. A. Scott, and T. J. R. Hughes. Isogeometric boundary-element analysis for the wave-resistance problem using T-splines. *Computer Methods in Applied Mechanics and Engineering*, submitted, 2014.

Saman Hosseini, Joris J.C. Remmers, Clemens V. Verhoosel, and René de Borst. An isogeometric continuum shell element for non-linear analysis. *Computer Methods in Applied Mechanics and Engineering*, 271:1 – 22, 2014.

T. J. R. Hughes. *The Finite Element Method: Linear Static and Dynamic Finite Element Analysis.* Dover Publications, Mineola, NY, 2000.

T. J. R. Hughes, J. A. Cottrell, and Y. Bazilevs. Isogeometric analysis: CAD, finite elements, NURBS, exact geometry, and mesh refinement. *Computer Methods in Applied Mechanics and Engineering*, 194:4135–4195, 2005.

Thomas J. R. Hughes, John A. Evans, and Alessandro Reali. Finite element and NURBS approximations of eigenvalue, boundary-value, and initial-value problems. *Computer Methods in Applied Mechanics and Engineering*, 272:290 – 320, 2014.

X. Li and M. A. Scott. Analysis-suitable T-splines: characterization, refineability, and approximation. *Mathematical Models and Methods in Applied Science*, 24(06):1141–1164, 2014.

X. Li, J. Zheng, T. W. Sederberg, T. J. R. Hughes, and M. A. Scott. On linear independence of T-spline blending functions. *Computer Aided Geometric Design*, 29:63 – 76, 2012.

Lei Liu, Yongjie Zhang, Thomas J. R. Hughes, Michael A. Scott, and Thomas W. Sederberg. Volumetric T-spline construction using boolean operations. In Josep Sarrate and Matthew Staten, editors, *Proceedings of the 22nd International Meshing Roundtable*, pages 405–424. Springer International Publishing, 2014.

D. Schillinger, L. Dedé, M. A. Scott, J. A. Evans, M. J. Borden, E. Rank, and T. J. R. Hughes. An isogeometric design-through-analysis methodology based on adaptive hierarchical refinement of NURBS, immersed boundary methods, and T-spline CAD surfaces. *Computer Methods in Applied Mechanics and Engineering*, 249 – 252:116 – 150, 2012.

Robert Schmidt, Roland Wüchner, and Kai-Uwe Bletzinger. Isogeometric analysis of trimmed NURBS geometries. *Computer Methods in Applied Mechanics and Engineering*, 241–244:93 – 111, 2012.

M. A. Scott. *T-splines as a Design-Through-Analysis technology*. PhD thesis, The University of Texas at Austin, 2011.

M. A. Scott, M. J. Borden, C. V. Verhoosel, T. W. Sederberg, and T. J. R. Hughes. Isogeometric Finite Element Data Structures based on Bézier Extraction of T-splines. *International Journal for Numerical Methods in Engineering,*, 88:126 – 156, 2011.

M. A. Scott, X. Li, T. W. Sederberg, and T. J. R. Hughes. Local refinement of analysis-suitable T-splines. *Computer Methods in Applied Mechanics and Engineering*, 213:206 – 222, 2012.

M. A. Scott, R. N. Simpson, J. A. Evans, S. Lipton, S. P. A. Bordas, T. J. R. Hughes, and T. W. Sederberg. Isogeometric boundary element analysis using unstructured T-splines. *Computer Methods in Applied Mechanics and Engineering*, 254:197 – 221, 2013.

M. A. Scott, D. C. Thomas, and E. J. Evans. Isogeometric spline forests. *Computer Methods in Applied Mechanics and Engineering*, 269:222 – 264, 2014.

T. W. Sederberg, J. Zheng, A. Bakenov, and A. Nasri. T-splines and T-NURCCs. *ACM Trans. Graph.*, 22:477–484, July 2003.

T. W. Sederberg, D. L. Cardon, G. T. Finnigan, N. S. North, J. Zheng, and T. Lyche. T-spline simplification and local refinement. *ACM Trans. Graph.*, 23:276–283, August 2004.

T. W. Sederberg, G. T. Finnigan, X. Li, and H. Lin. Watertight trimmed NURBS. *ACM Trans. Graph.*, 27:1–79, August 2008.

R. N. Simpson, M. A. Scott, M. Taus, D. C. Thomas, and H. Lian. Acoustic isogeometric boundary element analysis. *Computer Methods in Applied Mechanics and Engineering*, 269:265–290, 2014.

T-Splines, Inc. http://www.tsplines.com/products/t-tools-libraries.html, 2011.

Derek C. Thomas, Michael A. Scott, John A. Evans, Kevin Tew, and Emily J. Evans. Bézier projection: a unified approach for local projection and quadrature-free refinement and coarsening of NURBS and T-splines with particular application to isogeometric design and analysis. *Computer Methods in Applied Mechanics and Engineering*, (to appear), 2014.

C. V. Verhoosel, M. A. Scott, R. de Borst, and T. J. R. Hughes. An isogeometric approach to cohesive zone modeling. *International Journal for Numerical Methods in Engineering,*, 87:336 – 360, 2011a.

C. V. Verhoosel, M. A. Scott, T. J. R. Hughes, and R. de Borst. An isogeometric analysis approach to gradient damage models. *International Journal for Numerical Methods in Engineering,*, 86:115–134, 2011b.

W. A. Wall, M. A. Frenzel, and C. Cyron. Isogeometric structural shape optimization. *Computer Methods in Applied Mechanics and Engineering*, 197:2976–2988, 2008.

W. Wang, Y. Zhang, M. A. Scott, and T. J. R. Hughes. Converting an unstructured quadrilateral mesh to a standard T-spline surface. *Computational Mechanics,*, 48:477 – 498, 2011.

GPSR Compliance

The European Union's (EU) General Product Safety Regulation (GPSR) is a set of rules that requires consumer products to be safe and our obligations to ensure this.

If you have any concerns about our products, you can contact us on ProductSafety@springernature.com

In case Publisher is established outside the EU, the EU authorized representative is:

Springer Nature Customer Service Center GmbH
Europaplatz 3
69115 Heidelberg, Germany

Batch number: 10683837

Printed by Printforce, the Netherlands